Laboratory Manual

Human Biology

Fourteenth Edition

Sylvia S. Mader

Mc
Graw
Hill
Education

HUMAN BIOLOGY LABORATORY MANUAL, FOURTEENTH EDITION

Published by McGraw-Hill Education, 2 Penn Plaza, New York, NY 10121. Copyright © 2016 by McGraw-Hill Education. All rights reserved. Printed in the United States of America. Previous editions © 2014, 2012 and 2010. No part of this publication may be reproduced or distributed in any form or by any means, or stored in a database or retrieval system, without the prior written consent of McGraw-Hill Education, including, but not limited to, in any network or other electronic storage or transmission, or broadcast for distance learning.

Some ancillaries, including electronic and print components, may not be available to customers outside the United States.

This book is printed on acid-free paper.

1 2 3 4 5 6 7 8 9 0 RMN/RMN 1 0 9 8 7 6 5

ISBN 978-1-259-29300-9
MHID 1-259-29300-9

Senior Vice President, Products & Markets: *Kurt L. Strand*
Vice President, General Manager, Products & Markets: *Marty Lange*
Vice President, Content Design & Delivery: *Kimberly Meriwether David*
Managing Director: *Michael S. Hackett*
Brand Manager: *Chris Loewenberg*
Director, Product Development: *Rose Koos*
Senior Product Developer: *Anne Winch*
Digital Product Analyst: *Christine Carlson*
Marketing Manager: *Christine Ho*
Director, Content Design & Delivery: *Linda Avenarius*
Program Manager: *Angie Fitzpatrick*
Content Project Manager: *Mary Jane Lampe*
Buyer: *Sandy Ludovissy*
Design: *David W. Hash*
Content Licensing Specialists: *Lori Hancock, Leonard Behnke*
Cover Image: *Paragliding © Getty Images RF*
Art Studio and Compositor: *MPS North America LLC*
Typeface: *11/13 StixGeneral*
Printer: *R. R. Donnelley*

All credits appearing on page or at the end of the book are considered to be an extension of the copyright page.

The Internet addresses listed in the text were accurate at the time of publication. The inclusion of a website does not indicate an endorsement by the authors or McGraw-Hill Education, and McGraw-Hill Education does not guarantee the accuracy of the information presented at these sites.

www.mhhe.com

Contents

Preface   iv
To the Instructor   iv
To the Student   v

Lab	Title	Page	Text Chapter Reference
1	Scientific Method	1	1
2	Measuring with Metric	9	3
3	Light Microscopy	19	3
4	Chemical Composition of Cells	35	2
5	Cell Structure and Function	49	3
6	Body Tissues	61	4
7	Organization of the Body	75	4
8	Cardiovascular System	91	5, 6
9	Chemical Aspects of Digestion	105	9
10	Energy Requirements and Ideal Weight	115	9
11	Homeostasis	131	9, 10, 11
12	Musculoskeletal System	149	12, 13
13	Nervous System and Senses	167	14, 15
14	Reproduction and Development	183	17, 18
15	Mitosis and Meiosis	199	19
16	Patterns of Genetic Inheritance	217	21
17	DNA and Biotechnology	231	22
18	Human Evolution	247	23
19	Effects of Pollution on Ecosystems	265	24, 25

Appendix: Metric System   A-1
Credits   C-1
Index   I-1
Practical Examination Answer Sheet online at http://connect.mheducation.com

Applications for Daily Living

The Stress Test   7
Microscopic Diagnoses   32
Nutrition Labels   47
Dehydration and Water Intoxication   59
Tissue Engineering   73
The Systems Work Together   89
High Blood Pressure   103
Losing Weight   129

The Liver Is a Vital Organ   147
Bone Marrow Transplants   165
LASIK Surgery   181
Cord Around Baby's Neck   198
Mitosis and Cancer   214
Choosing the Gender of Your Child   229
Personal DNA Sequencing   245
Climate Change   272

Preface

To the Instructor

The laboratory exercises in this manual are coordinated with *Human Biology,* a text that has two primary functions: (1) to achieve an understanding of how the human body works, and (2) to show the relationship of humans to other living things in the biosphere.

This manual may also be used in coordination with other biology texts. There are a sufficient number of laboratories and exercises within each lab to tailor the laboratory experience as desired. Then, too, many exercises may be performed as demonstrations rather than as student activities, thereby shortening the time required to cover a particular concept.

The Exercises

All exercises have been tested for student interest, preparation time, time of completion, and feasibility. The following features are particularly appreciated by adopters:

Integrated Opening. Each laboratory begins with a list of Learning Outcomes organized according to the major sections of the laboratory. The major sections of the laboratory are numbered in the laboratory text material.

Prelab Questions. The Learning Outcomes contain one prelab question per major section. These questions are meant to encourage students to read the laboratory before they start work. Instructors can require that the prelab questions be answered and turned in on the day of the lab and/or instructors can select a few to include in a laboratory quiz.

Self-Contained Content. Each laboratory gives all the background information necessary to understand the concepts being studied and to answer the questions asked. This feature will reduce student frustration and increase learning.

Human Emphasis. This manual accompanies *Human Biology,* so it is fitting for the laboratories to have a human emphasis. This emphasis means that it is not necessary to include "Human" in the title of the laboratories because they all have a human orientation.

Scientific Process. This manual provides opportunities for students to gain an appreciation of the scientific method. The first laboratory of the manual explicitly explains the steps of the scientific method and gives students an opportunity to use them.

Student Activities. Sequentially numbered steps guide students as they perform each activity. Color bars are used to designate whether an activity is an Observation or whether it is an Experimental Procedure. A boxed statement with a clock icon is added at the start of the day's work to call attention to any exercises that require more time than other exercises. A clock also appears within the procedure whenever time is required for the results to occur.

Application for Daily Living. Relevancy is of primary importance in the *Human Biology Lab Manual.* Relevancy is particularly apparent due to a feature called *Application for Daily Living.* These short boxes at the end of the lab tell students how the day's work applies in a practical way to a current health or environmental concern.

Laboratory Safety. Laboratory safety is of prime importance, and the listing now online at **http://connect.mheducation .com**, will assist instructors in making the laboratory experience a safe one. Warning notices occur throughout the manual as is appropriate to the particular procedure.

The Fourteenth Edition

Most laboratories were revised this edition to improve clarity and student participation. In particular I would like to call your attention to these laboratories:

Laboratory 2: Measuring with Metric
This laboratory introduces students to the metric system in a way that engages students and leads to their understanding of and ability to use the metric system.

Laboratory 11: Homeostasis
This laboratory was reorganized so that the urinary system and the contribution of the kidneys to homeostasis is considered first. This arrangement better allows instructors to concentrate on the anatomy and physiology of the kidneys if they wish to do so. The contribution of the lungs and liver to homeostasis are also included as in the previous edition.

Laboratory 15: Mitosis and Meiosis
Karyotyping was removed from this laboratory and placed in laboratory 16, allowing this laboratory to concentrate on the normal process of meiosis. Other than the movement of karyotyping to laboratory 16, this laboratory remains as it was in the previous edition.

Laboratory 16: Patterns of Genetic Inheritance
New to this edition, the basics of blood typing has been added as an example of multiple allele inheritance.

Using simulated blood typing, students do an exercise that involves paternity testing. The last section of this lab, now entitled Genetic Counseling, proceeds from an examination of abnormal karyotypes to pedigree analysis and construction.

Laboratory 18: Human Evolution

This lab was rewritten and the illustrations were revised to improve the human emphasis. A new exercise was provided on extinct hominids.

Customized Editions

The 19 laboratories in this manual are now available as individual "lab separates," so instructors can custom-tailor the manual to their particular course needs. Find more information at **http://create.mheducation.com.**

LearnSmart® Labs

McGraw-Hill Education LearnSmart® Labs are available where noted to accompany many labs in this manual. These outcomes-based lab simulations assess a student's knowledge and adaptively correct deficiencies, allowing the student to learn faster and retain more knowledge with greater success. A demonstration of the labs can be found at **www.learnsmartadvantage.com.**

Laboratory Resource Guide

The *Laboratory Resource Guide,* an essential aid for instructors and laboratory assistants, free to adopters of the *Laboratory Manual,* is online at **http://connect.mheducation.com.** The answers to all the laboratory questions are in the *Resource Guide.*

To the Student

Special care has been taken in preparing *Human Biology Laboratory Manual* so that you will **enjoy** the laboratory experience as you **learn** from it. The instructions and discussions are written clearly so that you can understand the concepts presented and the procedures you are doing. Student learning aids discussed next are designed to help you focus on important aspects of each exercise.

Student Learning Aids

Students will want to be aware of these learning aids:

Learning Outcomes set the goals for each laboratory session. The major sections of each laboratory are numbered, and the Learning Outcomes are grouped according to these topics. The Learning Outcomes will help you review the material for a laboratory practical or any other kind of exam.

Prelab Questions are a part of the Learning Outcomes; when you can answer the Prelab Questions, you are prepared for a meaningful laboratory experience. These questions will guide your study of the concepts and the student activities in a meaningful way.

The Introduction on the same page as the Learning Outcomes reviews background information necessary for comprehending the upcoming experiments.

Color bars call out exercises that require your active participation. Different colors are used to highlight Observations and Experimental Procedures. If you see a clock in an experimental procedure, it means that a certain amount of time is required before you can observe your results.

Icons are used to get your attention. A safety icon tells you when to be particularly careful about the use of a chemical; a clock icon at the start of the day's work tells you when you might want to set up a procedure ahead of time.

Tables and fill-in-the-blanks allow you to record the results of investigations and experiments. Be sure to answer each question carefully. These allow you to understand the procedure. In addition, each laboratory ends with a set of review questions covering the day's work.

Laboratory Preparation

Read the Learning Outcomes and be sure to answer the Prelab Questions. Also *study* the introductory material and the procedures before the day you have lab. To obtain an even better understanding of a topic, consult your text.

Explanations and Conclusions

Throughout a laboratory, you are often asked to formulate explanations or conclusions. To do so, you will need to synthesize information from a variety of sources, including the following:

1. Your experimental results and/or the results of other groups in the class. If your data are different from those of other groups in your class, do not erase your answer; add the other answers in parentheses.
2. Your knowledge of underlying principles. Obtain this information from the laboratory Introduction or the appropriate section of the laboratory and from the corresponding chapter of your text.
3. Your understanding of how the experiment was conducted and/or the materials used. *Note:* Ingredients can be contaminated or procedures incorrectly followed, resulting in reactions that seem inappropriate. If this occurs, consult with other students and your instructor to see if you should repeat the experiment.

In the end, be sure you are truly writing an explanation or conclusion and not just restating the observations made.

Color Bar and Icon

Throughout each laboratory, a color bar designates either an Observation or an Experimental Procedure.

Observation: An activity in which models, slides, and preserved or live organisms are observed to achieve a learning outcome.

Experimental Procedure: An activity in which a series of steps uses laboratory equipment to gather data and come to a conclusion.

Time: An icon is used to designate when time is needed before you can determine your results. A boxed statement at the beginning of the laboratory calls attention to any experimental procedure requiring a considerable amount of time. Start these activities at the beginning of the laboratory, proceed to other activities, and return to this experimental procedure when appropriate.

Safety: An icon is used to designate when particular caution is needed. Your health and safety are of utmost concern, and you should follow the directive.

Laboratory Review

Each laboratory ends with a number of short-answer questions that will help you determine whether you have accomplished the outcomes for the laboratory.

Student Feedback

If you have any suggestions for how this laboratory manual could be improved, you can send your comments to

The McGraw-Hill Companies
Product Development—General Biology
501 Bell St.
Dubuque, IA 52001

Acknowledgments

We gratefully acknowledge the help of Natalia A. Abramova, University of Maine–Augusta in the development of this edition of the lab manual.

1

Scientific Method

Learning Outcomes

1.1 Using the Scientific Method
- Outline the steps of the scientific method.
- Distinguish among observations, hypotheses, conclusions, and theories.

Prelab Question: Explain what a scientist means by the term *theory*.

1.2 Observations Concerning Heart Rate
- Observe and describe the rhythmic beating of the heart.
- Read the data presented by the laboratory manual about the heartbeat.

Prelab Question: How do scientists carry out the first step of the scientific method?

1.3 Formulating Hypotheses
- Formulate a hypothesis based on the observations.

Prelab Question: Why is a hypothesis called a tentative explanation?

1.4 Performing an Experiment and Coming to a Conclusion
- Design an experiment that can be repeated by others.
- Reach a conclusion based on observation and experimentation.

Prelab Question: In science, a conclusion is based on the data. Explain.

Application for Daily Living: **The Stress Test**

Introduction

In everyday life we are often called upon to make observations and use our past experiences to come to a hypothesis that can be tested. For example, suppose you flipped the switch, but the light didn't come on. Most likely you would hypothesize that the bulb has burnt out, the socket isn't working, or there has been a power outage. Each of these possibilities could be tested until you have come to a conclusion that allows the problem to be fixed. In today's laboratory, we want to experience firsthand the manner in which a scientist uses these same steps to come to a conclusion.

1.1 Using the Scientific Method

Science differs from other human ways of knowing and learning by its process, which can be varied because it can be adjusted to where and how a study is being conducted. Still, the scientific process often involves the use of the scientific method. In today's laboratory, we will use the scientific method to come to a conclusion about the heartbeat rate. As depicted in Figure 1.1, the scientific method involves making observations, formulating hypotheses, doing experiments, and coming to a conclusion. Many conclusions on the same topic allow scientists to develop a scientific theory.

Figure 1.1 Flow diagram for the scientific method.
On the basis of observations, a scientist formulates a hypothesis. The hypothesis is tested by further experiments and/or observations. The scientist then concludes whether the results support or do not support the hypothesis. The return arrow shows that scientists often choose to retest the same hypothesis or test a related hypothesis. Conclusions from many different but related experiments may lead to the development of a scientific theory.

Observation

New observations are made, and previous data are studied.

Hypothesis

Input from various sources is used to formulate a testable statement.

Experiment/Observations

The hypothesis is tested by experiment or further observations.

Conclusion

The results are analyzed, and the hypothesis is supported or rejected.

Scientific Theory

Many experiments and observations support a theory.

Making observations: As a first step in their study of a topic, scientists use all their senses and also instruments, such as a microscope, to make **observations.** Observations can also include something you read, such as the information given on pages 3 and 4 about the beating of the heart. If you want to learn more about the heartbeat rate, you could do a search of the literature or talk to a nurse or doctor about the heartbeat. Scientists keep notes about their observations, as you will be asked to do. Why is it helpful for a scientist to begin by making observations? _____

_____

Formulating a hypothesis: Based on their observations, scientists come to a tentative explanation, called a **hypothesis,** about what it is they are investigating. Having observed the beating of the heart and having done a little literary research, you will be asked to formulate a hypothesis about the influence of exercise on the heartbeat rate. Why is a hypothesis sometimes called an "educated guess"?

_____

Testing your hypothesis: After formulating a hypothesis, a scientist performs an **experiment,** which tests the hypothesis. A well-designed experiment must have a **control,** a sample or event that is not exposed to the testing procedure. If the control and the test produce the same results, either the procedure is flawed or the hypothesis is wrong. What is the purpose of a control? _____

_____

The results of your experiment are called **data.** Data are any factual information that come to light because of your experiment or perhaps further observations you have made. It is important for a scientist to keep accurate records of all their data. When another person repeats the experiment, their data should be the same or something is amiss. Why must a scientist keep complete records of an experiment?

_____

Coming to a conclusion: The data will either support or not support the hypothesis. A scientist can come to a **conclusion** that the hypothesis has been shown to be false but does not say that a hypothesis has been shown to be true. After all, some fault might be found with the method of collecting data. Why don't scientists say they have proven their hypothesis true? _____

_____

Arriving at a scientific theory: A **theory** in science is an encompassing conclusion based on many individual conclusions in the same field. For example, the cell theory states that all organisms are composed of cells. The cell theory is based on observations made by many scientists over many years who have observed a countless variety of living things. We will not be developing any theories today. How is a scientific theory different from a conclusion? _____

_____

1.2 Observations Concerning Heart Rate

You will follow the procedure for the scientific method outlined in Figure 1.1. The first step in this procedure is to make observations about the topic you are proposing to study.

Observation: The Human Heart

1. Your instructor will show a short video of the heart beating. After viewing the video and studying Figure 1.2, do your observations suggest that the human heartbeat sends blood rhythmically

   into the blood vessels of the cardiovascular system? _____

2. In his text *Functional Human Anatomy,* James E. Crouch writes

   The heart is a pump which is of primary importance in the maintenance of the flow and pressure of the blood in the closed cardiovascular system. Its contractions start in the early days of embryonic development and must continue for

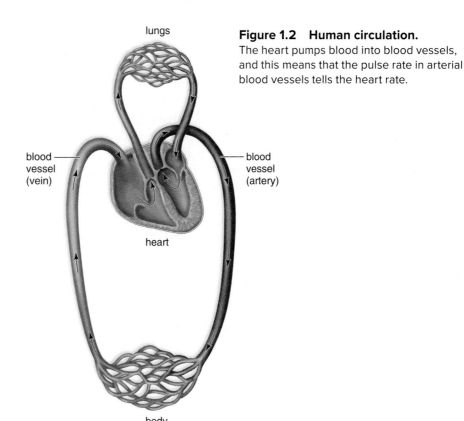

lungs

blood vessel (vein)

blood vessel (artery)

heart

body

Figure 1.2   Human circulation.
The heart pumps blood into blood vessels, and this means that the pulse rate in arterial blood vessels tells the heart rate.

the duration of the individual's life. For the contractions to cease even for a few minutes, would cause serious damage, or would be fatal. The heart beat occurs at an average rate of 72/min (72 times per minute) and add up to about 100,000 times a day or about 2,600,000,000 beats in a lifetime of seventy years. If you clench and open your fist alternately at the rate of 72/min, the muscles involved will "feel" tired after about 2 minutes....[1]

3. When you take a spinning or kickboxing class, your instructor may ask you to periodically take your pulse rate. The pulse rate indicates how fast the heart is beating. In your experience, how does exercise affect the heart rate and

   pulse rate? _____

1.3 Formulating Hypotheses

Now that you have made some observations of the beating heart, had some literary input, and recalled past experiences, it is time to formulate a hypothesis about the heart rate and exercise.

1. Think of an exercise a subject could do in the laboratory at a slow speed and a fast speed.

   Record your choice here: _____

2. After a class discussion and input from your instructor, the class will have the subject do the same type exercise for the same length of time during the experiment. Together the class will decide what is meant by a "slow speed" and "fast speed" exercise and the time limit for each. Three minutes of exercise is suggested. Record the class decision about type of slow/fast exercise to be done and for how long:

   Slow exercise/time: **Fast exercise/time:**

   _____   _____

3. Formulate a hypothesis here that tells how you expect the heart rate to be affected by the chosen slow and fast exercise per length of time:

   Hypothesis: The heart rate will _____

   _____

1.4 Performing an Experiment and Coming to a Conclusion

Variables

1. Variables are differences in your experimental procedure or results. Two variables of value are the experimental variable and the dependent variable. The **experimental variable** is what you purposely change to get your results. What is your experimental variable?

   _____

   The **dependent variables** are the variables that the hypothesis predicts will be influenced by the experimental variable. What will be a dependent variable in your experiment?

   _____

2. The experimenter tries to control any other variables that could affect the results. For example, if a scientist were doing an experiment with plants, s/he would always use the same type plant and the same type pots and would subject the plants to the same treatment under the same circumstances. **Controlled variables** are aspects of an experiment the scientist makes consistent to prevent them from affecting the results. Otherwise, the experiment has too many variables and the results might be due to differences in treatment other than the one being tested. What are some possible variables that should be controlled or they could affect the reliability of your results? _____

   _____

[1]Crouch, James E. 1985. *Functional Human Anatomy,* 4th ed. Philadelphia. Lea & Febiger, p. 399.

3. Subject variability is often a concern when doing experiments with humans. Just as all the people signed up for an exercise class are different (Fig. 1.3), so are the people taking a human biology course. Then, too, we have to consider that some subjects are used to doing exercise and some are not. The hope is that if subjects are randomly chosen, and the results are averaged, then differences between subjects may cancel themselves out. We also have to realize that in general the larger the sampling size, the more confidence there is in the results. The Framingham heart study, on which much of our common knowledge regarding heart disease is now based, began with 5,209 subjects. The best we can do under the present circumstances is to pool our results as a class while realizing that even this sample size is probably inadequate. Skill of the experimenter is another variable that we have to consider, and all students should practice taking the pulse rate in the manner described next.

Figure 1.3 An aerobics class.
Just as the individuals taking spinning class are varied in their appearance, so participants in a scientific study are different in their physiological characteristics.

Experimental Design

1. It's time to consider what the experimenter is to do. The experimenters will take pulse rate measurements at the wrist using the radial artery, which pulses every time the heart beats.
   a. **Location of the Radial Artery.** The radial artery is located on the thumb side of the wrist, a little below the base of the thumb. If you are right-handed, use the tips of your index and third finger of your right hand to find the radial artery on your left wrist. (Left-handed people should do the opposite.) You should be able to feel the pulsing of the artery.
   b. **Taking a measurement.** Choose a timepiece to use, preferably a stopwatch, but if this is not available, use a watch with a second hand as a way to tell the number of seconds. To take a measurement, use only the tips of your index and third finger. Holding gently, begin counting the pulse with "zero," and count for 10 seconds. Multiply by six, and the answer is your pulse rate per minute. Record your results:

   counts per 10 seconds _____ × 6 = _____ pulse rate per minute.

   c. When taking the pulse rate after exercising, begin as soon as the subject stops. If you let time pass, the pulse rate will slow, and your count will not be accurate.
2. As mentioned, every experiment needs to have a control. For this experiment, (a) the control should be a student who does not experience exercise. However, (b) the experiment could use a student who does exercise as long as the student fully recovers in between slow and fast exercise. Why is the first method better than the second

   method of providing a control? _____

   _____

   _____

   To generate enough data, the experiment will use the second method and assume that pulse rate before slow exercise and again, after recovery and before fast exercise, is the control sample (see Table 1.2).

Table 1.1	Physical Fitness Evaluation Chart for the Canadian Home Fitness Test	
	Ten-Second Pulse Rate	
Age (yr)	After First 3 Minutes of Exercise	After Second 3 Minutes of Exercise
15–19	If 30 or more, stop exercising.	If 26 or less, you have the recommended personal fitness level.
20–29	If 29 or more, stop exercising.	If 25 or less, you have the recommended personal fitness level.
30–39	If 28 or more, stop exercising.	If 24 or less, you have the recommended personal fitness level.
40–49	If 26 or more, stop exercising.	If 23 or less, you have the recommended personal fitness level.
50–59	If 25 or more, stop exercising.	If 22 or less, you have the recommended personal fitness level.
60–69	If 24 or more, stop exercising.	If 22 or less, you have the recommended personal fitness level.

3. Table 1.1 gives the 10-second pulse rate that tells various ages when they should stop exercising. Use the following grid to create a line graph using the data from the first and second columns of Table 1.1. *Label the x-axis (horizontal) "Age of subject." Use only the first age of each category. Label the y-axis (vertical) "Maximum 10 sec pulse rate/3 mins exercise."*

The graph tells us that as age _____, the acceptable 10 sec pulse rate after 3 mins of exercise _____.

4. As scientists do, before you begin your experiment, summarize your experimental design. Include information concerning the subjects; the experimental variable (what the subjects will do); and how the results, called the data, will be gathered.

The Experiment

1. The members of your class will work in groups. Each group will contain at least one experimenter and one subject. You may switch off and be the experimenter for one round of the experiment and the subject for another round of the experiment. Record your results in Table 1.2

Table 1.2 Increase in Heart Rate Following Slow and Fast Exercise per Group

	Pulse Rate/min Before Exercise	Pulse Rate/min Slow Exercise	Pulse Rate/min Before Exercise	Pulse Rate/min Fast Exercise
First subject				
Second subject				

2. Your instructor or designated members of the class will collect the data, average the results, and tell you what data to record in Table 1.3.

Table 1.3 Increase in Heart Rate Following Slow and Fast Exercise per Class

	Pulse Rate/min Before Exercise	Pulse Rate/min Slow Exercise	Pulse Rate/min Before Exercise	Pulse Rate/min Fast Exercise
Class average				

Conclusion: Heart Rate Following Exercise

- Did the results of your experiment support the hypothesis, per your group? _____
  per class? _____ Why do you say so? _____
  _____

Application for Daily Living

The Stress Test

A high pulse rate could possibly be an indication of heart irregularities. But a cardiac stress test, which examines a series of ECGs as the patient exercises (Fig. 1.4), is more reliable than taking the pulse rate after exercise. An abnormal ECG, which measures the electrical activity of the heart, is suggestive of an impending heart attack. One out of every five deaths in the United States is due to a heart attack, and there are over 1 million new and recurrent heart attacks per year. Heart attack is a fairly common occurrence, so it is often a good idea to determine the chances of having one.

During a stress test, the level of exercise is increased in 3-minute stages of ever-increasing incline and speed. The American Heart Association recommends a stress test as the first choice for patients with medium risk of coronary heart disease based on risk factors, such as smoking and family history of cardiac disorders. Over the past 20 years, more modern methods have been devised to test whether the heart is functioning properly, but they are more expensive and more invasive than a stress test.

Figure 1.4 Cardiac stress test.

_____   1. Which is more comprehensive, a conclusion or a theory?

_____   2. What is a tentative explanation of observed phenomena?

_____   3. What do you call the information scientists collect when doing experiments and making observations?

_____   4. What step in the scientific method follows experiments and observations?

_____   5. What do you call a sample that goes through all the steps of an experiment, but is not exposed to the experimental variable?

Indicate whether statements 6 and 7 are hypotheses, conclusions, or scientific theories.

_____   6. The data show that vaccines protect people from disease.

_____   7. All living things are made of cells.

_____   8. The heart rate is the same as what other cardiovascular system rate?

_____   9. What is the function of a beating heart?

_____   10. What is the normal heart rate when at rest?

_____   11. What is your evidence that a beating heart is critical to good health?

Thought Questions

12. To save time, Xavier decides to have Jose do the slow exercise test and Maria the fast exercise test. Criticize Xavier's procedure.

13. How does the method you use to decide what to wear differ from that of the scientific method?

Find even more resources to learn the chapter concepts including animations, tutorial videos, and interactive practice questions.

McGraw-Hill Access Science Website

An online encyclopedia of science and technology that provides information, including videos, that can enhance the laboratory experience.

www.accessscience.com

The following LearnSmart Labs contain exercises that are related to the content of this chapter:

• Scientific Method

2

Measuring with Metric

Learning Outcomes

2.1 Length
- Compare and contrast the metric units for length: meter (m), centimeter (cm), millimeter (mm), micrometer (μm), and nanometer (nm).
- Know the abbreviations for each of these units.
- Convert the metric units for length from one type of unit to another.

Prelab Question: Is nanometer, centimeter, or micrometer the larger measurement unit for length?

2.2 Weight
- Compare and contrast the metric units for weight: kilogram (kg), gram (g), and milligram (mg).
- Know the abbreviations for each of these units.
- Convert the metric units for weight from one type of unit to another.

Prelab Question: Would you expect the weight of a penny to be given in terms of kilograms or grams?

2.3 Volume
- Compare and contrast the metric units for volume: liter (l) and mliliter (ml).
- Know the abbreviations for each of these units.
- Convert the metric units for volume from one type of unit to another.

Prelab Question: What units of measurement appear on a graduated cylinder?

2.4 Temperature
- Compare and contrast the Fahrenheit (F) and Celsius (C) temperature scales.
- Know the abbreviations for each of these units.
- Convert one type temperature into another using a provided equation.

Prelab Question: Water boils at 100° in which scale?

Introduction

The **metric system** is the standard system of measurement in the sciences, including biology, chemistry, and physics (Fig. 2.1). It has tremendous advantages because all conversions, whether for volume, mass (weight), or length, are in units of ten. This base-ten system is similar to our monetary system, in which 10 cents equals a dime, 10 dimes equals a dollar, and so on. In this laboratory, you will get experience making measurements of length, volume, mass, and also temperature.

Figure 2.1
The metric system is the system of measurement used in scientific laboratories.

2.1 Length

Metric units of length measurement include the **meter (m), centimeter (cm), millimeter (mm), micrometer (μm),** and **nanometer (nm)** (Table 2.1).

Table 2.1 Metric Units of Length Measurement				
Unit	Meters	Centimeters	Millimeters	Relative Size
Meter (m)	1 m	100 cm	1,000 mm	Largest
Centimeter (cm)	0.01 (10^{-2}) m	1 cm	10 mm	
Millimeter (mm)	0.001 (10^{-3}) m	0.1 cm	1.0 mm	
Micrometer (μm)	0.000001 (10^{-6}) m	0.0001 (10^{-4}) cm	0.001 (10^{-3}) mm	
Nanometer (nm)	0.000000001 (10^{-9}) m	0.0000001 (10^{-7}) cm	0.000001 (10^{-6}) mm	Smallest

You will want to know what these abbreviations stand for, so write them out here:

m = _meter_   μm = _micrometer_

cm = _centimeter_   nm = _nanometer_

mm = _millimeter_

How many cm are in a meter? _100_   How many mm are in a centimeter? _10_

How many μm are in a millimeter? _1000_   How many nm are in a micrometer? _10000_

Meter, Centimeter, and Millimeter

Observation: A Meterstick

1. Obtain a meterstick. On one side, find the numbers 1 through 39, which denote inches. One meter equals 39.37 inches; therefore, 1 meter is roughly equivalent to 1 yard.

Figure 2.2   The end of a meterstick.

2. Turn the meterstick over and observe the metric subdivisions (Fig. 2.2). How many centimeters are in a meter? _100_ The prefix *centi-* means 100. For example, how many cents are in a dollar? _100_

3. How many millimeters are in a centimeter? _10_ But the prefix *milli-* means a thousand. How many millimeters are in a meter? _1000_ Obtain a penny and measure its width in terms of mm. _19mm_ Why does it seem preferable to measure a penny in terms of mm? _Doesn't reach whole #._ _gives Smaller measurement_

4. Use the meterstick and the method shown in Figure 2.3 to measure the length of two long bones from a disarticulated human skeleton. Lay the meterstick flat on the lab table. Place a long bone next to the meterstick between two pieces of cardboard (each about 10 cm × 30 cm), held upright at

right angles to the stick. The narrow end of each piece of cardboard should touch the meterstick. The length between the cards is the length of the bone in centimeters. For example, if the bone measures from the 22 cm mark to the 50 cm mark, the length of the bone is __28__ cm. If the bone measures from the 22 cm mark to midway between the 50 cm and 51 cm marks, its length is __28.5__ cm = __285__ mm.

5. Record the length of two bones. First bone: __43__ cm = __430__ mm.

   Second bone: _____ cm = _____ mm.

Figure 2.3 Measurement of a long bone.
How to measure a long bone using a meterstick.

Millimeter, Micrometers, and Nanometer

As you will discover in the next laboratory, the units micrometer (μm) and nanometer (nm) are useful in microscopy because the objects being viewed under the compound light microscope are cells or the contents of cells. Figure 2.4 shows that cells are generally smaller than a millimeter (mm).

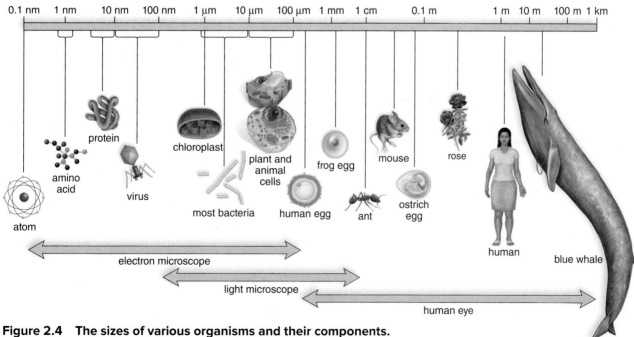

Figure 2.4 The sizes of various organisms and their components.

1. Obtain a small metric ruler marked in centimeters (cm) and millimeters (mm). Micrometers (μm) are not on the ruler, and it is necessary to remember that there are 1,000 μm in a mm. Use the ruler to measure the diameter of the circle shown below to the nearest mm.

   Diameter of circle in mm: _____38_____ mm

2. Use this formula to convert this measurement to μm. Do you expect the answer to be a smaller number or a larger number? ____Larger____

$$\underline{38} \ mm \times \frac{1,000 \ \mu m}{mm} = \underline{38000} \ \mu m$$

   Size of circle in μm: __38,000__ μm

3. Knowing that there are 1,000 nm in a μm, change the formula to convert these μm into nm.

   38,000 × 1,000 = 38,000,000 nm

   Diameter of circle in nm: __38,000,000__ nm

4. You have shown that the diameter of this circle is __38__ mm = __38,000__ μm = __38,000,000__ nm.

   It is expected when doing microscopy that you will make such conversions.

2.2 Weight

For our purposes we will equate mass with weight. Just as a meter is the basic unit of length, the **kilogram** is the basic unit of mass.

> 1 kilogram (kg) =1,000 **gram** (g)
> 1g = 1,000 **milligrams** (mg)

The weight of domestic animals or a human would be given in kilograms but usually biologists work with the units of grams or milligrams. Using a revision of the formula on page 12 if necessary, do these conversions:

2 g = _____ mg and 0.2 g = _____ mg.

Experimental Procedure: Weight

Examine the balance scales available in your laboratory. Originally balance scales had two pans. The object to be weighed was placed in one pan and weights are placed on the other pan. When the pans contained exactly the same mass the beam between them was in balance. Today digital scales contain internal weighing mechanisms and an electronic display shows the weight (Fig. 2.5).

Figure 2.5   A digital scale.

1. Use a digital or balance scale to measure the weight of a wooden block small enough to hold in the palm of your hand. Weigh the block to a tenth of a gram. The weight of the wooden block is

   _____ g = _____ mg.

2. Measure the weight of an item small enough to fit inside the opening of a 50 ml graduated cylinder. The

   item, a(n) _____, is _____ g = _____ mg.

3. A triple beam balance gets its name from its three horizontal beams. If you are going to use a triple beam balance, study it closely (Fig. 2.6). Each beam has a moveable weight. Move the weights in order to see that the closest beam weighs to 0.1 g; the middle beam has 100 g graduations, and the farthest beam has 10 g graduations.

   Before making any measurements, clean the weighing pan and move all the weights to the far left. The weights should line up to indicate zero grams; if they do not, turn the adjustment knob until they do.

a.

b.

Figure 2.6   A triple beam balance.
a. The entire balance. **b.** Enlargement of triple (3) beams.

Measure the weight of an object by placing it in the center of the weighing pan and moving the weights until the beams balance (are straight across). The weight of the object is the sum of the weights on the three beams.

If so directed by your instructor, use a triple beam balance to take the weight of one or more of these objects to a tenth of a gram:

Penny   _____

Paper clip   _____

Quarter   _____

2.3   Volume

Two metric units of volume are the **liter (l)** and the **milliliter (ml).** One liter = 1,000 ml.

Experimental Procedure: Volume

1. Volume measurements can be related to those of length. For example, use a millimeter ruler to measure the wooden block used in the previous Experimental Procedure to get its length, width, and depth.

   height

   length = _____ cm; width = _____ cm; depth = _____ cm

   The volume, or space, occupied by the wooden block can be expressed in cubic centimeters

   (cc or cm³) by multiplying: length × width × depth = _____ cm³. For purposes of this

   Experimental Procedure, 1 cubic centimeter equals 1 milliliter; therefore, the wooden block has

   a volume of = _____ ml.

2. In the biology laboratory, liquid volume is usually measured directly in liters or milliliters with appropriate measuring devices. For example, use a 50 ml graduated cylinder to add 20 ml of water to a test tube. First, fill the graduated cylinder to the 20 ml mark. To do this properly, you have to make sure that the lowest

Figure 2.7 Meniscus.
The proper way to view the meniscus.

meniscus reading 20 ml

improper position

proper position

improper position

margin of the water level, or the **meniscus** (Fig. 2.7), is at the 20 ml mark. Place your eye directly parallel to the level of the meniscus, and add water until the meniscus is at the 20 ml mark. (Having a dropper bottle filled with water on hand can help you do this.) A large, blank, white index card held behind the cylinder can also help you see the scale more clearly. Now pour the 20 ml of water into the test tube.

3. Hypothesize how you could find the total volume of the test tube. _____

_____

_____

What is the test tube's total volume? _____

4. Fill a 50 ml graduated cylinder with water to about the 20 ml mark. Hypothesize how you could use this setup to calculate the volume of the small object you weighed previously (see step 2, p. 13). _____

_____

Now perform the operation you suggested. The object, _____, has a volume of _____ml.

5. Hypothesize how you could determine how many drops from the pipet of the dropper bottle equal 1 ml. _____

How many drops from the pipet of the dropper bottle equal 1 ml? _____

6. Some pipets are graduated (Fig. 2.8) and can be filled to a certain level as a way to measure volume directly. Your instructor will demonstrate this. Are pipets customarily used to measure large or small volumes? _____

Bulb

Figure 2.8 Use of a graduated pipet.
Squeezing the bulb allows solutions to enter the pipet. Replace the bulb with your index finger to hold the liquid in place while transferring it to a different container.

2.4 Temperature

There are two temperature scales: the **Fahrenheit (F)** and **Celsius (centigrade, C)** scales (Fig. 2.9). Scientists use the Celsius scale.

Experimental Procedure: Temperature

1. Study the two scales in Figure 2.9, and complete the following information:

   a. Water freezes at _32_ °F = _0_ °C.

   b. Water boils at _212_ °F = _100_ °C.

2. To convert from the Fahrenheit to the Celsius scale, use the following equation:

$$°C = (°F - 32°)/1.8$$

or

$$°F = (1.8°C) + 32$$

Human body temperature of 98°F is what temperature on the Celsius scale? _37_

3. Record any two of the following temperatures in your lab environment. In each case, allow the bulb end of the Celsius thermometer to remain in or on the sample for one minute.

   Room temperature = _17_ °C

   Surface of your skin = _____ °C

   Cold tap water in a 50 ml beaker = _____ °C

   Hot tap water in a 50 ml beaker = _59_ °C

   Ice water = _0_ °C

Figure 2.9 Temperature scales.
The Fahrenheit (°F) scale is on the left, and the Celsius (°C) scale is on the right.

Laboratory Review 2

_kg_   1. Would kg, ml, or cm signify a measurement of weight?

grad. Cylinder   2. Would volume be measured with a meterstick, a graduated cylinder, or a scale?

micrometer   3. What does the abbreviation μm stand for?

_mm_   4. What is the smallest unit of measurement shown on a meterstick?

meters

5. Would you expect the height of a ceiling to be in meters, millimeters, or cubic centimeters?

_111 g_

6. If a triple beam balance shows a weight of 100 g plus 10 g plus 1 g, what is the weight of the object?

_1100 μm_

7. Convert 1.1 mm to μm.

_0.5 L_

8. Convert 500 ml to l.

_5,000 mg_

9. Convert 5 g to mg.

_15,000 μm_

10. Convert 1.5 cm to μm.

$1 cm = 10,000 μm$

Thought Questions

11. If an object weighs .001 g, it also weighs how many mg?

_1 mg_

12. If an object is 100 mm in length, it is also how many cm in length?

_10 cm_

13. An object is added to a graduated cylinder that holds 250 ml, and the water rises to 300 ml. What is the volume of the object in ml? in cm^3?

_50 ml_

14. Name two units of measurement you expect to use in the next laboratory, which concerns microscopy.

Nano

Micro

3

Light Microscopy

Learning Outcomes

3.1 Light Microscopes Versus Electron Microscopes
- Describe three differences between the compound light microscope and the electron microscope.

Prelab Question: Greater magnification enlarges the subject, but what is another equally important attribute of a microscope? Explain.

3.2 Stereomicroscope (Binocular Dissecting Microscope)
- Identify the parts of and tell how to focus the stereomicroscope.

Prelab Question: If you wanted to examine the external features of a fly, you might decide to use a stereomicroscope. Why?

3.3 Use of the Compound Light Microscope
- Name and give the function of the basic parts of the compound light microscope.
- List the steps for bringing an object into focus with the compound light microscope.
- Describe how the image is inverted by the compound light microscope.
- Calculate the total magnification and the diameter of field of view for both low- and high-power lens systems.
- Explain how a slide of colored threads provides information on the depth of field.

Prelab Question: What's the difference between low power and high power when using the compound light microscope?

3.4 Microscopic Observations
- Name and describe three types of cells studied in this exercise.

Prelab Question: What is there about live *Euglena* that makes them difficult to study microscopically?

Application for Daily Living: **Microscopic Diagnoses**

Introduction

This laboratory introduces you to the features, functions, and use of the stereomicroscope and the compound light microscope. Transmission and scanning electron microscopes are explained, and micrographs produced by using these microscopes are shown throughout these exercises. The stereomicroscope and the scanning electron microscope view the surface and/or the three-dimensional structure of an object. The compound light microscope and the transmission electron microscope can view only extremely thin sections of a specimen. If a subject was sectioned lengthwise for viewing, the interior of the projections at the top of

a. Cell   **b.** Longitudinal section   **c.** Cross section

Figure 3.1   Longitudinal and cross sections.
a. Transparent view of a cell. **b.** A longitudinal section would show the cilia at the top of the cell. **c.** A cross section shows only the interior where the cut is made.

the cell, called cilia, would appear in the micrograph (Fig. 3.1*b*). A lengthwise cut through any type of specimen is called a **longitudinal section** (**ls**). On the other hand, if the subject was sectioned crosswise below the area of the cilia, you would see other portions of the interior of the subject (Fig. 3.1*c*). A crosswise cut through any type of specimen is called a **cross section** (**cs**).

3.1 Light Microscopes Versus Electron Microscopes

Biological objects can be small, so we often use a microscope to view them. Many types of instruments, ranging from the hand lens to the electron microscope, are effective magnifying devices. A short description of two types of light microscopes and two types of electron microscopes follows.

Light Microscopes

Light microscopes use visible light rays magnified and focused by means of lenses. The **stereomicroscope** (binocular dissecting microscope) is designed to study entire objects in three dimensions at low magnification. The **compound light microscope** is used for examining small or thinly sliced sections of objects under higher magnification than that of the stereomicroscope. The term **compound** refers to the use of two sets of lenses: the ocular lenses located near the eyes and the objective lenses located near the object. Illumination is from below, and visible light passes through clear portions but does not pass through opaque portions. To improve contrast, the microscopist uses stains or dyes that bind to cellular structures and absorb light. Figure 3.2*a* is a **photomicrograph,** a photograph of an image produced by a compound light microscope.

Figure 3.2 Comparative micrographs.
Micrographs of a lymphocyte, a type of white blood cell. **a.** A photomicrograph (light micrograph) shows less detail than a **(b.)** transmission electron micrograph (TEM). **c.** A scanning electron micrograph (SEM) shows the cell surface in three dimensions.

a. Photomicrograph (LM) of a lymphocyte   2150×

b. Transmission electron micrograph (TEM)   2150×
of a lymphocyte

c. Scanning electron micrograph (SEM)   5000×
of a lymphocyte

Electron Microscopes

Electron microscopes use a beam of electrons magnified and focused on a photographic plate by means of electromagnets. The **transmission electron microscope** is analogous to the compound light microscope. For electron microscopy, the object is ultra-thinly sliced and treated with heavy metal salts to provide contrast. Figure 3.2*b* is a micrograph produced by this type of microscope. The **scanning electron microscope** is analogous to the dissecting light microscope. It gives an image of the surface and dimensions of an object, as is apparent from the scanning electron micrograph in Figure 3.2*c*.

The micrographs in Figure 3.2 demonstrate that much smaller objects can be viewed with electron microscopes than with compound light microscopes. The difference between these two types of microscopes, however, is not simply a matter of magnification; it is also the electron microscope's ability to show detail. The electron microscope has greater resolving power. **Resolution** is the minimum distance between two objects at which they can still be seen, or resolved, as two separate objects. The use of high-energy electrons rather than light gives electron microscopes a much greater resolving power because two objects much closer together can still be distinguished as separate points. Table 3.1 lists several other differences between the compound light microscope and the transmission electron microscope.

Table 3.1 Comparison of the Compound Light Microscope and the Transmission Electron Microscope

Compound Light Microscope	Transmission Electron Microscope
1. Glass lenses	1. Electromagnetic lenses
2. Illumination by visible light	2. Illumination due to a beam of electrons
3. Resolution $\cong$ 200 nm	3. Resolution $\cong$ 0.1 nm
4. Magnifies to 2,000×	4. Magnifies to 1,000,000×
5. Costs up to tens of thousands of dollars	5. Costs up to hundreds of thousands of dollars

3.2 Stereomicroscope (Binocular Dissecting Microscope)

The stereomicroscope (binocular dissecting microscope) allows you to view objects in three dimensions at low magnifications. It is used to study entire small organisms, any object requiring lower magnification, and opaque objects that can be viewed only by reflected light. It is also called a stereomicroscope because it produces a three-dimensional image.

Identifying the Parts

After your instructor has explained how to carry a microscope, obtain a stereomicroscope and a separate illuminator, if necessary, from the storage area. Place it securely on the table. Plug in the power cord, and turn on the illuminator. There is a wide variety of stereomicroscope styles, and your instructor will discuss the specific style(s) available to you. Regardless of style, the following features should be present:

1. **Binocular head:** Holds two eyepiece lenses that move to accommodate the various distances between different individuals' eyes.
2. **Eyepiece lenses:** The two lenses located on the binocular head. What is the magnification of your eyepieces? _____ Some models have one **independent focusing eyepiece** with a knurled knob to allow independent adjustment of each eye. The nonadjustable eyepiece is called the **fixed eyepiece.**

3. **Focusing knob:** A large black or gray knob located on the arm; used for changing the focus of both eyepieces together.
4. **Magnification changing knob:** A knob, often built into the binocular head, used to simultaneously change magnification in both eyepieces. This may be a **zoom** mechanism or a **rotating lens** mechanism of different powers that clicks into place.
5. **Illuminator:** Used to illuminate an object from above; may be built into the microscope or separate.

Locate each of these parts on your stereomicroscope, and *label them on Figure 3.3.*

Focusing the Stereomicroscope

1. Place a plastomount, that contains small organisms, in the center of the stage.
2. Adjust the distance between the eyepieces on the binocular head so that they comfortably fit the distance between your eyes. Using both eyes, you should be able to see an organism in the plastomount as one three-dimensional image.

Figure 3.3   Stereomicroscope.
Label this microscope with the help of the text material.

3. Use the focusing knob to bring an organism in the plastomount into focus.

4. Does your microscope have an independent focusing eyepiece? _yes_ If so, use the focusing knob to bring an organism in the plastomount in the fixed eyepiece into focus, while keeping the eye at the independent focusing eyepiece closed. Then adjust the independent focusing eyepiece so that the image is clear, while keeping the other eye closed. Is the image inverted? _yes_

5. Turn the magnification changing knob, and determine the type of mechanism on your microscope. A zoom mechanism allows continuous viewing while changing the magnification. A rotating lens mechanism blocks the view of the object as the new lenses are rotated. Be sure to click each lens firmly into place. If you do not, the field will be only partially visible. Which of these mechanisms is on your microscope? _All_

6. Set the magnification changing knob on the lowest magnification. *Choose an organism from the plastomount and sketch it in the following circle as though this represents your entire field of view:*

7. Rotate the magnification changing knob to the highest magnification. *Draw another circle within the one provided to indicate the reduction of the field of view.*
8. Experiment with various objects at various magnifications until you are comfortable with using the binocular dissecting microscope.
9. When you are finished, return your stereomicroscope and illuminator to their correct storage areas.

Conclusions: Microscopy

- Which two of the three types of microscopes studied (compound light microscope, transmission electron microscope, stereomicroscope) view the surface of an object? _____

- Which two types of microscopes view objects that have been sliced and treated to improve contrast? _____

- Of the three microscopes, which one resolves the greater amount of detail?

_____

3.3   Use of the Compound Light Microscope

As mentioned, the name **compound light microscope** indicates that it uses two sets of lenses and light to view an object. The two sets of lenses are the ocular lenses located near the eyes and the objective lenses located near the object. Illumination is from below, and the light passes through clear portions but does not pass through opaque portions. This microscope is used to examine small or thinly sliced sections of objects under higher magnification than would be possible with the stereomicroscope.

Identifying the Parts

Obtain a compound light microscope from the storage area, and place it securely on the table. Identify the following parts on your microscope, and *label them in Figure 3.4.*

1. **Eyepieces (ocular lenses):** What is the magnifying power of the ocular lenses on your microscope? _____10 x_____
2. **Body tube:** Holds nosepiece at one end and eyepiece at the other end; conducts light rays.
3. **Arm:** Supports upper parts and provides carrying handle.
4. **Nosepiece:** Revolving device that holds objectives.
5. **Objectives** (objective lenses):
   a. **Scanning power objective:** This is the shortest of the objective lenses and is used to scan the whole slide. The magnification is stamped on the housing of the lens. It is a number followed by a multiplication sign (×). What is the magnifying power of the scanning lens on your microscope? _____5 x_____

Figure 3.4 Compound light microscope.
Compound light microscope with binocular head and mechanical stage. Label this microscope with the help of the text.

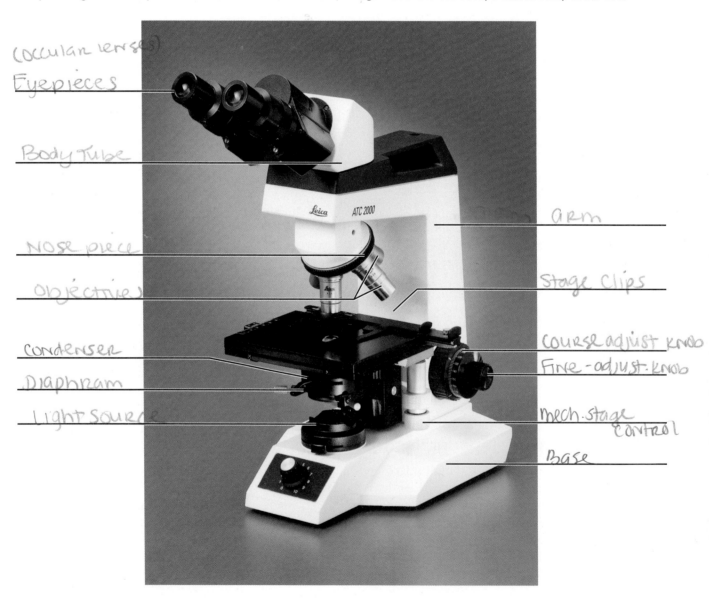

    b. **Low-power objective:** This lens is longer than the scanning lens and is used to view objects in greater detail. What is the magnifying power of the low-power objective lens on

        your microscope? _10x_____

    c. **High-power objective:** If your microscope has three objective lenses, this lens will be the longest. It is used to view an object in even greater detail. What is the magnifying power

        of the high-power objective lens on your microscope? _40x_____

    d. **Oil immersion objective** (on microscopes with four objective lenses): Holds a 95× (to 100×) lens and is used in conjunction with immersion oil to view objects with the greatest

        magnification. Does your microscope have an oil immersion objective? _100x – yes_

        If this lens is available, your instructor will discuss its use when the lens is needed.

6. **Coarse-adjustment knob:** Knob used to bring object into approximate focus; used only with low-power objective.
7. **Fine-adjustment knob:** Knob used to bring object into final focus.
8. **Condenser:** Lens system below the stage used to focus the beam of light on the object being viewed.
9. **Diaphragm** or **diaphragm control lever:** Controls amount of light used to view the object.
10. **Light source:** An attached lamp that directs a beam of light up through the object.
11. **Base:** The flat surface of the microscope that rests on the table.
12. **Stage:** Holds and supports microscope slides.
13. **Stage clips:** Hold slides in place on the stage.
14. **Mechanical stage** (optional): A movable stage that aids in the accurate positioning of the slide.

    Does your microscope have a mechanical stage? _____

15. **Mechanical stage control knobs** (optional): Two knobs usually located below the stage. One knob controls forward/reverse movement, and the other controls right/left movement.

Rules for Microscope Use

1. Figure 3.5 shows the proper way to carry the microscope.
2. The lowest power objective (scanning or low) should be in position both at the beginning and at the end of microscope use.
3. Use only lens paper for cleaning lenses.
4. Do not tilt the microscope as the eyepieces could fall out, or wet mounts could be ruined.

Figure 3.5 The proper way to carry a microscope.

5. Keep the stage clean and dry to prevent rust and corrosion.
6. Do not remove parts of the microscope.
7. Keep the microscope dust-free by covering it after use.
8. Report any malfunctions.
9. Do not use coarse focus when viewing a specimen with the high-power objective.

Focusing the Microscope—Lowest Power

1. Turn the nosepiece so that the *lowest*-power lens is in straight alignment over the stage.
2. Always begin focusing with the *lowest*-power objective lens (4× [scanning] or 10× [low power]).
3. With the coarse-adjustment knob, lower the stage (or raise the objectives) until it stops.
4. Place a slide of the letter *e* on the stage, and stabilize it with the clips. (If your microscope has a mechanical stage, pinch the spring of the slide arms on the stage, and insert the slide.) Center the *e* as best you can on the stage or use the two control knobs located below the stage (if your microscope has a mechanical stage) to center the *e*.
5. Again, be sure that the lowest-power objective is in place. Then, as you look from the side, decrease the distance between the stage and the tip of the objective lens until the lens comes to an automatic stop or is no closer than 3 mm above the slide.
6. While looking into the eyepiece, rotate the diaphragm (or diaphragm control lever) to give the maximum amount of light.
7. Using the coarse-adjustment knob, slowly increase the distance between the stage and the objective lens until the object—in this case, the letter *e*—comes into view, or focus.
8. Once the object is seen, you may need to adjust the amount of light. To increase or decrease the contrast, slightly rotate the diaphragm.
9. Use the fine-adjustment knob to sharpen the focus if necessary.
10. Practice having both eyes open when looking through the eyepiece, as this greatly reduces eyestrain.

Inversion

Inversion refers to a microscopic image that is upside down and reversed.

Observation: Inversion

1. Draw the letter *e* as it appears on the slide (with the unaided eye, not looking through the eyepiece). _____
2. Draw the letter *e* as it appears when you look through the eyepiece. _____
3. What differences do you notice? _____
4. Move the slide to the right. Which way does the image appear to move? _____
   Explain. _____

Focusing the Microscope—Higher Powers

Compound light microscopes are **parfocal**—once the object is in focus with the lowest power, it should also be almost in focus with the higher power.

1. Bring the object into focus under the lowest power by following the instructions in the previous section.
2. Make sure that the letter *e* is centered in the field of the lowest objective.
3. Move to the next higher objective (low power [10×] or high power [40×]) by turning the nosepiece until you hear it click into place. Do not change the focus; parfocal microscope objectives will not hit normal slides when changing the focus if the lowest objective is initially in focus. (If you are on low power [10×], proceed to high power [40×] before going on to step 4.)
4. If any adjustment is needed, use only the *fine*-adjustment knob. (Note: Always use only the fine-adjustment knob and not the coarse-adjustment knob with high power.)
5. On a drawing of the letter *e, draw a circle around the portion of the letter that you are now seeing with high-power magnification.* _____
6. When you have finished your observations of this slide (or any slide), rotate the nosepiece until the lowest-power objective clicks into place, lower the stage, and then remove the slide.

Total Magnification

Total magnification is calculated by multiplying the magnification of the ocular lens (eyepiece) by the magnification of the objective lens.

Observation: Total Magnification

Calculate total magnification figures for your microscope, and record your findings in Table 3.2.

Table 3.2	Total Magnification			
Objective		Ocular Lens	Objective Lens	Total Magnification
Scanning power (if present)		×	×	×
Low power		×	×	×
High power		×	×	×

Field of View

A microscope's field of view is the circle visible through the lenses. The diameter of field is the length of the field from one edge to the other. Like any other measurement in science, the diameter of the field is measured using metric units. The metric system has tremendous advantages because all conversions are in units of ten. Table 3.3 provides the metric units for length. The basic unit of length is the meter (m), and Table 3.3 tells you how the other units of length relate to the meter. You would use a meter to measure your car or this classroom. Examine a meterstick on display. It is divided into centimeters (cm) and millimeters (mm). Look at 1 mm, and imagine dividing up that amount of space into 1,000 micrometers (μm). A meter is much too large a unit for microscopy. The units appropriate for microscopic use are highlighted in Table 3.3. They are

1.0 millimeter (mm) = 1,000 micrometers (μm) = 1,000,000 nanometers (nm)

For light microscopy, the most useful unit of these is μm, and you will be asked to convert millimeters into micrometers by multiplying by 1,000 in the exercises that follow.

Table 3.3	Metric Unit Values					
Unit	Symbol	Value (m)	Value (cm)	Value (mm)	Value (μm)	Value (nm)
Meter	m	1.0 m	100 cm	1,000 mm	1,000,000 μm	1,000,000,000 nm
Centimeter	cm	0.01 m	1.0 cm	10 mm	10,000 μm	10,000,000 nm
Millimeter	mm	0.001 m	0.1 cm	1.0 mm	1,000 μm	1,000,000 nm
Micrometer	μm	0.000001 m	0.0001 cm	0.001 mm	1.0 μm	1,000 nm
Nanometer	nm	0.000000001 m	0.0000001 cm	0.000001 mm	0.001 μm	1.0 nm

Low-Power (10×) Diameter of Field

1. Place a clear, plastic metric ruler across the stage so that the edge of the ruler is visible as a horizontal line along the diameter of the low-power (not scanning) field. Be sure that you are looking at the millimeter side of the ruler.

2. Estimate the number of millimeters, to tenths, that you see along the field: _____ mm.

   (*Hint:* Start with one of the millimeter markers at the edge of the field.) Convert the number to micrometers: _____ μm. This is the **low-power diameter of field** (**lpd**) for your microscope in micrometers.

High-Power (40×) Diameter of Field

1. To compute the **high-power diameter of field** (**hpd**), substitute these data into the formula given:

   a. lpd = low-power diameter of field (in micrometers)   = _____ μm

   b. lpm = low-power total magnification (from Table 3.2) = _____ ×

   c. hpm = high-power total magnification (from Table 3.2) = _____ ×

$$hpd = lpd \times \frac{lpm}{hpm}$$

$$hpd = (\quad μm) \times \frac{(\quad)}{(\quad)} = \text{_____} μm$$

Conclusion: Total Magnification and Field of View

- What unit of metric measurement is most useful for light microscopy? _____

  Use the micrometer row in Table 3.3 to convert 1 μm to millimeters. 1 μm = _____ mm

  Use the meter row to convert 1 m to μm. 1 m = _____ μm

- Does low power or high power have a larger field of view (allowing you to see more of the object)?

  _____

- Does low power or high power have a smaller field but magnifies to a greater extent? _____

- To locate small objects on a slide, first find them under _____; then place them in the

  center of the field before rotating to _____.

Depth of Field

When viewing an object on a slide under high power, the **depth of field** (Fig. 3.6) is the area—from top to bottom—that comes into focus while slowly focusing up with the microscope's fine-adjustment knob.

Observation: Depth of Field

1. Obtain a prepared slide with three or four colored threads mounted together, or prepare a wet-mount slide with three or four crossing threads or hairs of different colors. (Directions for preparing a wet mount are given in Section 3.4.)

2. With low power, find a point where the threads or hairs cross. Slowly focus up and down. Notice that when one thread or hair is in focus, the others seem blurred. Determine the order of the threads or hairs, and

complete Table 3.4. Remember, as the stage moves upward (or the objectives more downward), objects on top come into focus first.

Table 3.4 Order of Threads (or Hairs)	
Depth	Thread (or Hair) Color
Top	
Middle	
Bottom	

3. Switch to high power, and notice that the depth of field is more shallow with high power than with low power. Constant use of the fine-adjustment knob when viewing a slide with high power will give you an idea of the specimen's three-dimensional form. For example, viewing a number of sections allows reconstruction of the three-dimensional structure, as demonstrated in Figure 3.6.

Figure 3.6   Depth of field.
A demonstration of how focusing at depths 1, 2, and 3 would produce three different images (views) that could be used to reconstruct the original three-dimensional structure of the object.

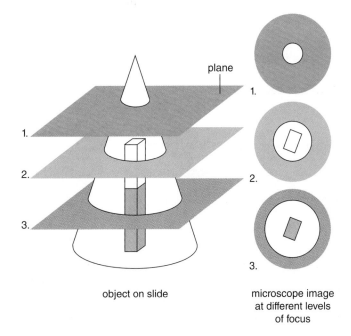

object on slide

microscope image at different levels of focus

3.4 Microscopic Observations

In this part of the laboratory session, you will have an opportunity to first examine a prepared slide and then two slides you will prepare yourself.

Human Blood Cells

Human blood contains red blood cells that transport oxygen about the body and white blood cells that fight infection. You will have no difficulty finding the red blood cells on the slide because they are numerous (see Fig. 3.2a).

1. Obtain a prepared slide of human blood, and mount it on the microscope stage. Use the method you learned previously to bring the red blood cells into focus. Practice observing different parts of the slide and switching to high power. Can you find a white blood cell? White blood cells are generally much larger than red blood cells.

2. Red blood cells are small, only 7–8 μm in diameter, and their small size can account for why there are so many in the blood. Blood is a liquid tissue, and this gives us an opportunity to refer to the liquid measurement in science. The metric system is uniform. Table 3.3 can be used to describe the units of liquid measurement, as long as you appropriately change the terms.

Instead of these terms: Use:

meter (m)   liter (l)
centimeter (cm)   centiliter (cl)
millimeter (mm)   milliliter (ml)
micrometer (μm)   microliter (μl)

There are generally 4–6 million red blood cells in a microliter (μl). To appreciate this quantity, observe the micropipette on display. If so advised by your instructor, use the micropipette to measure a μl and disperse this quantity into a beaker.

Wet Mounts

When a specimen is prepared for observation, the object should always be viewed as a **wet mount.** A wet mount is prepared by placing a drop of liquid on a slide or, if the material is dry, by placing it directly on the slide and adding a drop of water or stain. The mount is then covered with a coverslip, as illustrated in Figure 3.7. Dry the bottom of your slide before placing it on the stage.

b. Lower coverslip slowly.

a. Add drop of suspension or dry object and solution.

Figure 3.7   Preparation of a wet mount.

Human Epithelial Cells

Epithelial cells cover the body's surface and line its cavities.

Observation: Human Epithelial Cells

1. Obtain a prepared slide, or make your own as follows:

   a. Obtain a prepackaged flat toothpick (or sanitize one with alcohol or alcohol swabs).

   b. Gently scrape the inside of your cheek with the toothpick, and place the scrapings on a clean, dry slide. Discard used toothpicks in the biohazard waste container provided.

   c. Add a drop of very weak methylene blue or iodine solution, and cover with a coverslip.

2. Observe under the microscope.

3. Locate the nucleus (the central, round body), the cytoplasm (interior), and the plasma membrane (outer cell boundary). *Label Figure 3.8.*

4. Your epithelial slides are biohazardous, so they must be disposed of as indicated by your instructor.

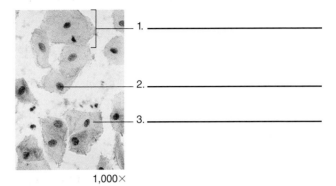

1. _____

2. _____

3. _____

1,000×

Figure 3.8   Cheek epithelial cells.
Label the nucleus, the cytoplasm, and the plasma membrane.

Euglena

Examination of *Euglena* (a unicellular organism with a flagellum to facilitate movement) will test your ability to observe objects with the microscope, to use depth of field, and to control illumination to heighten contrast.

Observation: Euglena

1. Make a wet mount of *Euglena* by using a drop of a *Euglena* culture and adding a drop of Protoslo (methyl cellulose solution) onto a slide. The Protoslo slows the organism's swimming.

2. Mix thoroughly with a toothpick, and add a coverslip.

3. Scan the slide for *Euglena:* Start at the upper left-hand corner, and move the slide forward and back as you work across the slide from left to right. The *Euglena* may be at the edge of the slide because they show an aversion to Protoslo. Use Figure 3.9 to help identify *Euglena.*

4. Experiment by using scanning, low-power, and high-power objective lenses; by focusing up and down with the fine-adjustment knob; and by adjusting the light so that it is not too bright.

5. Compare your *Euglena* specimens with Figure 3.9. Can you distinguish between the anterior and posterior ends? (The anterior end has a red eye spot.) _____ Does *Euglena* move forward or backward? _____

Does *Euglena* have internal structures? _____

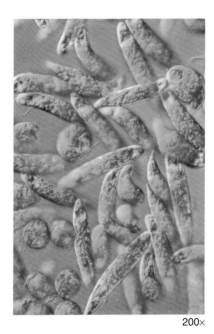

200×

Figure 3.9 Euglena.
Euglena is a unicellular, flagellated organism.

Microscopic Diagnoses

Microscopic examination of discharges, tissues, and the blood plays an important part of a physical exam. For example, your doctor may order a differential blood cell count when he or she believes it will help diagnose your illness. Or, he or she may microscopically examine a discharge if you may have a particular bacterial sexually transmitted disease (STD).

The Pap test, which women have during a routine gynecological exam, is a low-cost, easy microscopic examination of cells taken from the cervix located at the entrance to the womb. Pap tests look for abnormal cells in the lining of the cervix to determine if they are precancerous or cancerous. The more abnormal the appearance of the cells, the greater the concern (Fig. 3.10).

To do a Pap test, a physician merely takes a sample of cells from the cervix, which is then microscopically examined for signs of abnormality. Pap tests are credited with preventing over 90% of deaths from cervical cancer.

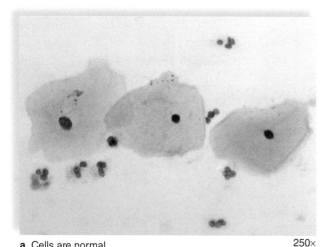

a. Cells are normal.   250×

b. Cells are precancerous.   250×

c. Cells are cancerous.   250×

Figure 3.10 Pap test.
A Pap test is a microscopic examination of cells from the cervix at the entrance to the womb. **a.** The cells are normal and cancer is not present. **b.** The cells are precancerous. **c.** The cells are abnormal and cancer is present.

scanning   **1.** Of the three types of microscopes studied, which one best shows the surface of an object?

_____   **2.** The resolving power of a light microscope is lower than that of an _____ microscope.

2 lenses   **3.** What does *compound* refer to in the designation *compound light microscope?*

_____   **4.** Which type of microscope would you use to view a *Euglena* swimming in pond water?

_____   **5.** What are the ocular lenses?

_____   **6.** Which objective always should be in place when beginning to use the microscope and also when putting it away?

_____   **7.** A total magnification of 100× requires the use of the 10× ocular lens with which of the objective lenses?

_____   **8.** Which part of a microscope controls the amount of light?

_____   **9.** What word is used to indicate that if the object is in focus at low power it will also be in focus at high power?

_____   **10.** If the thread layers are red, brown, green, from top to bottom, which layer will come into focus first if you are using the microscope properly?

_____   **11.** What adjustment knob is used with high power?

_____   **12.** If *Euglena* are swimming to the left, which way should you move your slide to keep them in view?

_____   **13.** What is the final item placed on a wet mount before viewing with a light microscope?

3d objects   **14.** What type of object do you study with a stereomicroscope?
Living specimens

Thought Questions

15. Why is a stereomicroscope also called a binocular dissecting microscope?

16. What is the advantage of focusing on different levels of an object?

|BIOLOGY

Find even more resources to learn the chapter concepts including animations, tutorial videos, and interactive practice questions.

McGraw-Hill Access Science Website

An online encyclopedia of science and technology that provides information, including videos, that can enhance the laboratory experience.

www.accessscience.com

The following LearnSmart Labs contain exercises that are related to the content of this chapter:

• Microscopy Biology

4

Chemical Composition of Cells

Benedicts=heat

Introduction

The U.S. Food and Drug Administration is charged with the task of ensuring that our foods are safe and wholesome. Investigators and inspectors visit more than 15,000 facilities a year, overseeing that food products are not contaminated and are labeled truthfully. About 3,000 products a year are found to be unfit for consumers and are withdrawn from the marketplace, so the task of testing foods is a necessary one.

Food products contain three major components: proteins, carbohydrates, and lipids. Today you will learn how to test a food for the presence of these components. You will also learn how proteins, carbohydrates, and lipids are constructed and, therefore, why they give a positive result with particular reagents. A positive result indicates that a component is present, and a negative result indicates that a component is not present.

What is a Control?

The experiments in today's laboratory have both a positive control and a negative control. The **positive control** goes through all the steps of the experiment and does contain the substance being tested. Therefore, positive results are expected. The **negative control** goes through all the steps of the experiment except it does not contain the substance being tested. Therefore, negative results are expected.

For example, if a test tube contains glucose (the substance being tested) and Benedict's reagent (blue) is added, a red color develops upon heating. This test tube is the positive control; it tests positive for glucose. If a test tube does not contain glucose and Benedict's reagent is added, Benedict's is expected to remain blue. This test tube is the negative control; it tests negative for glucose.

What benefit is a positive control? Positive controls give you a standard by which to tell if the substance being tested is present (or acting properly) in an unknown sample. Negative controls ensure that the experiment is giving reliable results; after all, if a negative control should happen to give a positive result, then the entire experiment may be faulty and cannot be relied on.

4.1 Carbohydrates Benedicts (heat)

Carbohydrates include starch, glycogen, and sugars. Starch and sugar are major components of baked goods, such as breads, cakes, and cookies. Flour contains starch, but starch does not taste sweet. The disaccharide, called **sucrose,** which does taste sweet, is often added to baked goods. Although we enjoy the sweet taste of sucrose, research has shown it is detrimental to our health. Researchers at the Harvard School of Public Health found that subjects who consumed sugar-sweetened sodas more than once per day for eight years exhibited an 80% increased risk of the illness diabetes mellitus. Also, they gained 17 pounds on the average, showing the link between sugar consumption and obesity. Obese individuals tend to suffer from cardiovascular disease more than those of normal weight.

> **Planning Ahead** Heating the water bath in advance for the test for sugars (page 39) may save valuable time.

Starch and Glycogen

When you digest your food, the starch component from plants is broken down to maltose and then maltose is digested to **glucose,** the sugar found in your blood (Fig. 4.1).

maltose $C_{12}H_{22}O_{11}$ water   glucose $C_6H_{12}O_6$ glucose $C_6H_{12}O_6$

disaccharide   +   water   ⇌   monosaccharide   +   monosaccharide

Figure 4.1 Maltose, a disaccharide.
During a digestion reaction, the components of water are added, and the bond is broken. During a synthesis reaction, a bond forms between the two glucose molecules, the components of water are removed, and maltose results.

Your body stores excess glucose as glycogen, often called animal starch, in the liver. Both glycogen and starch are long chains of glucose molecules; therefore, they are **polysaccharides** (Fig. 4.2).

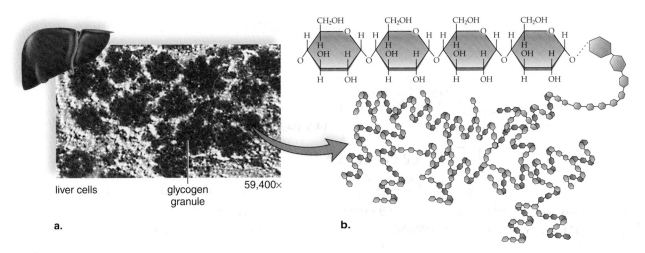

Figure 4.2 Glycogen and starch.
Glycogen is a polysaccharide composed of many glucose units as is starch. **a.** Photomicrograph of glycogen granules in liver cells. **b.** Structure of glycogen. Glycogen is a very branched molecule, while starch is not branched.

liver cells

glycogen granule

59,400×

a.

b.

Test for Starch *iodine - only tells if starch is present*

Many foods contain starch. The test for starch is to add an iodine solution. If the substance contains starch, the color shifts from brown to a deep purple to black, as noted in Table 4.1. This positive result is explained thus: Iodine molecules lodge in the coiled structure of starch. This causes a change in the way that iodine molecules are able to reflect light and the color we observe.

Table 4.1 Positive Test for Starch	
	Starch
Iodine solution (brown)	Black

Experimental Procedure: Test for Starch

In this procedure, you will be running a test for starch much as an employee of the Food and Drug Administration would do. Starch, as you know, is present in a number of different baked goods.

Before running the tests, answer these questions.

1. Choose which of these substances would be in a test tube if you detected starch: starch, albumin, water, iodine. _____

2. Which substances listed would you include in a negative-control test tube? _____

   Why? _____

3. Why might you run another test using albumin in a test tube instead of a food expected to contain starch? _____

The Procedure

1. Label five clean test tubes (1 to 5).
2. Using the designated graduated transfer pipets, add 1 ml of the experimental solutions listed in Table 4.2 to the test tubes according to their numbers.
3. Then add five drops of iodine solution to the tubes at the same time.
4. Note the final color changes and record your observations in Table 4.2

Iodine test for starch

iodine = gold

Table 4.2 Iodine (IKI) Tests for Starch

Tube	Contents	Color	Conclusions
1	Starch suspension	Cloudy white (Black)	POSITIVE
2	Water	Clear (Gold)	NO REACTION
3	Potato juice	Light Brown (Rose Gold)	POSITIVE
4	Albumin solution	Yellow Peach (Gold)	NO-REACTION
5	onion	Cloudy white (yellow)	NO REACTION
6	glucose	Clear (light yellow)	NO Reaction

Conclusions: Test for Starch

- From your test results, draw conclusions about whether starch is present in each tube. Write these conclusions in Table 4.2. Explain your data and conclusions. _____

_____

- Based on your results, is iodine a valid test for starch? _____

Explain why. _____

Experimental Procedure: Microscopic Study of Potato

1. With a sharp razor blade, carefully slice a very thin piece of potato. Place it on a microscope slide, add a drop of water and a coverslip, and observe under low power with your compound light microscope. Compare your slide with the photomicrograph of starch granules (Fig. 4.3). Find the cell wall (large, geometric compartments) and the starch grains (numerous clear, oval-shaped objects).

2. Without removing the coverslip, place two drops of iodine solution onto the microscope slide so that the iodine touches the coverslip. Draw the iodine under the coverslip by placing a small piece of paper towel in contact with the water on the **opposite** side of the coverslip.

3. Microscopically examine the potato again on the side closest to where the iodine solution was applied.

What is the color of the small oval bodies?

Light Brown

What is the chemical composition of these oval bodies?

_____

starch granule

57×

Figure 4.3 Starch.
Photomicrograph of starch granules in potato cells.

Test for Sugars

Monosaccharides and some disaccharides will react with **Benedict's reagent** after being heated in a boiling water bath. The color change can range from green to red, and increasing concentrations of sugar will give a continuum of colored products (Table 4.3). This experiment tests for the presence (or absence) of varying amounts of sugars in a variety of materials and chemicals.

Table 4.3 Benedict's Reagent (Some Typical Reactions)

Chemical	Chemical Category	Benedict's Reagent (After Heating)
Water	Inorganic	Blue (no change)
Glucose	Monosaccharide (carbohydrate)	Varies with concentration: very low—green low—yellow moderate—yellow-orange high—orange very high—orange-red
Maltose	Disaccharide (carbohydrate)	Varies with concentration—see "Glucose"
Starch	Polysaccharide (carbohydrate)	Blue (no change)

Experimental Procedure: Test for Sugars

1. Prepare a boiling water bath and label five clean test tubes (1 to 5).
2. Using the designated graduated transfer pipets, add 1 ml of the experimental solutions listed in Table 4.4 to the test tubes according to their numbers.
3. Then add five drops of Benedict's reagent to all the tubes at this time.
4. Place all the tubes into the boiling water bath at the same time.
5. When, after a few minutes, you see a change of colors, remove all the tubes from the water bath and record your observations in Table 4.4.
6. Save your tubes for comparison purposes when you do Section 4.4.

> ⚠ **Benedict's reagent** Benedict's reagent is highly corrosive. Exercise care in using this chemical. If any should spill on your skin, wash the area with mild soap and water. Follow your instructor's directions for disposal of this chemical.

Benedict's test for sugar

Table 4.4 Benedict's Reagent Test

Tube	Contents	Initial	Color (After Heating)	Conclusions
1	Glucose solution	Lt. Blue	Orange	positive
2	Water	Blue	Blue	no Reaction
3	Onion juice	Lt. Blue	yellow/orange	positive
4	Starch suspension	Lt. Blue	sky Blue	no reaction
5	potato Juice	Brown	yellow	positive

Conclusions: Test for Sugar

- From your test results, conclude whether sugar is present and enter your conclusions in Table 4.4. Explain your data and conclusions. _____

_____

_____

- Based on your results, is Benedict's reagent a valid test for sugar? _____
  Explain. _____

_____

4.2 Proteins Biorets

Proteins have numerous functions in cells. Some are present for functional reasons, and others provide structural integrity.

Antibodies are functional proteins that combine with disease-causing pathogens as part of the body's immune response.

Transport proteins combine with and move substances from place to place. Hemoglobin packed inside red blood cells transports oxygen throughout the body. Albumin is another protein in our blood that performs several important roles including fatty acid transport.

Regulatory proteins control cellular metabolism in some way. For example, the hormone insulin regulates the amount of glucose in blood so that cells have a ready supply of energy.

Enzymatic proteins speed chemical reactions. A reaction that could take days or weeks to complete can happen within an instant if the correct enzyme is present. Amylase is an enzyme that speeds the breakdown of starch in the mouth and small intestine.

Structural proteins include keratin, found in hair, and myosin, found in muscle.

Hemoglobin is a protein. 1,740×

Amino Acids and Peptides

Proteins are long chains of **amino acids** joined together. About 20 different common amino acids are found in cells. All amino acids have an acidic group (—COOH) and an amino group (H₂N—), each linked to a central carbon by a separate covalent bond. Amino acids differ by the *R* group (rest of molecule) attached to the central carbon atom. The *R* groups have varying compositions, sizes, shapes, and chemical activities.

Hair is a protein.

Amino acid

A chain of two or more amino acids is called a **peptide,** and the bond between the two amino acids is called a **peptide bond** (Fig. 4.4). A **polypeptide** is a long chain of amino acids. A protein can be comprised of one or more polypeptides. When you digest your food, the **protein** component is broken down to amino acids, and then your body uses these amino acids to synthesize peptides and your proteins.

Figure 4.4 Peptide bond.
Peptide bond formation between amino acids creates a dipeptide. In a polypeptide, many amino acids are held together by multiple peptide bonds. During a digestion reaction, water is added, and the peptide bond is broken. The synthesis reaction involves the removal of one water molecule.

Test for Proteins

Biuret reagent (blue color) contains a strong solution of sodium or potassium hydroxide ($NaOH$ or KOH) and a small amount of dilute copper sulfate ($CuSO_4$) solution. The reagent changes color in the presence of proteins or peptides because the peptide bonds of the protein or peptide chemically combine with the copper ions in Biuret reagent. In this reaction, two positive results are possible as described in Table 4.5. The copper ions react with amino groups, deforming polypeptide chains and causing a resultant change in color.

> ⚠ **Biuret reagent** Biuret reagent is highly corrosive. Exercise care in using this chemical. If any should spill on your skin, wash the area with mild soap and water. Follow your instructor's directions for its disposal.

Table 4.5   Positive Test for Protein and Peptides

	Protein	Peptides
Biuret reagent (blue)	Purple	Pinkish-purple

If a positive result occurs and protein is present, what color will appear after using the Biuret test?

_____

If a negative result occurs and protein is not present, what color will remain after using the Biuret test?

_____

water = Neg. control

In this procedure you will be running a test for protein, much as an employee of the Food and Drug Administration would do. Egg white is composed of the protein called **albumin.** Therefore, albumin can be present in a number of different foods.

Before running the tests, answer these questions.

1. Which of these substances would be in the test tube if you detected protein? albumin, starch, water, Biuret reagent. _____

2. Which substances listed would you include in a negative-control test tube? _____
   Why? _____

3. Why might you run another test using starch in the test tube instead of the food expected to contain protein? _____

   If the results were negative, would it suggest that Biuret reagent is specific for protein? _____

The Procedure

1. Label three clean test tubes (1 to 3).
2. Using the designated graduated transfer pipets, add 1 ml of the experimental solutions listed in Table 4.6 to the test tubes according to their numbers.
3. Then add five drops of Biuret reagent to the tubes, with swirling to mix.
4. The reaction is almost immediate. Record your observations in Table 4.6.

Biuret test for protein and peptides

make note about initial color

Table 4.6	Biuret Test		
Tube	Contents	Final Color	Conclusions
1	Protein *Cloudy/pearl (Blue)*	*purple*	*positive*
2	Distilled water *Clear (Blue)*	*Blue tint (clear)*	*No reaction*
3	Starch suspension *Opaque/cloudy (Blue)*	*cloudy (clear)*	*No reaction*
4	*Pepsin oily yellow (Blue)*	*(Blue) Lavender*	*positive*

Conclusions: Test for Proteins

- From your test results, did you conclude that protein is present in any of the three test solutions? Enter your conclusions in Table 4.6. Explain your data and conclusions. *pepsin & protein were both positive reaction*

- Based on your results, is Biuret reagent a valid test for protein? _____
  Explain why. _____

4.3 Lipids

A number of different molecules, including fats and oils, are grouped together as **lipids** because they are insoluble in water. Fats and oils differ by their fatty acid content. A **fatty acid** consists of a long hydrocarbon chain and ends in an acid group (Fig. 4.5). The hydrocarbon chain can have bonds that differ like this:

| Saturated | Unsaturated | Trans fats |
| (butter) | (oils) | (hydrogenated oils) |

A saturated fatty acid has no double bonds between the carbon atoms, while an unsaturated fatty acid does have double bonds between certain of the carbon atoms. Saturated fatty acids are found in a solid fat, such as butter, whereas unsaturated fatty acids are found in liquid oils. They are liquid because the double bond creates a kink that prevents close packing of hydrocarbon chains. Saturated fats, but not unsaturated oils, encourage the development of plaque in the arteries of the body, leading to cardiovascular disease. Even more harmful than naturally occurring saturated fats are the so-called trans fats, found in vegetable oils that have been partially hydrogenated to make them more solid. Partial hydrogenation does not saturate all bonds. Instead, some hydrogen atoms end up on different sides of the chain. Processed and fast foods are apt to contain trans fats.

Fats and Oils

When you digest your food, fats and oils are broken down to one molecule of glycerol and three fatty acids (Fig. 4.5). Later, your body rejoins these molecules to form fat, which is a long-term energy source.

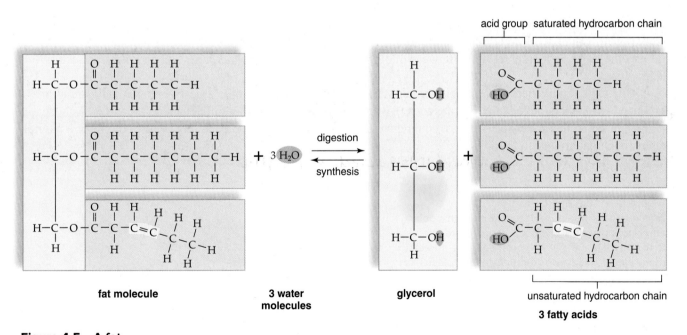

Figure 4.5 A fat.
During a digestion reaction, water is added, and the bonds are broken. A fat molecule forms when glycerol joins with three fatty acids as three water molecules are removed during a synthesis reaction.

Test for Fats and Oils

Fats and oils do not evaporate from brown paper; instead, they leave an oily spot.

Experimental Procedure: Test for Fats and Oils

1. Place a small drop of water on a square of brown paper. Describe the immediate effect. _Water_
   * _soaked into paper immediately_

2. Place a small drop of vegetable oil on a square of brown paper. Describe the immediate effect.
   oil is sitting on top of paper & absorbing @ slower rate

3. Wait at least 15 minutes for the paper to dry. Evaluate which substance penetrates the paper and which is subject to evaporation. Record your observations in Table 4.7.

Table 4.7	Test for Fats and Oils	
Sample	Observations	Conclusions
Water spot	Dried	No Reaction
Oil spot	Still sitting on surface	Positive

Emulsification of Lipids   milk = emulsification

Some molecules are **polar**, meaning that they have charged groups or atoms, and some are **nonpolar**, meaning that they have no charged groups or atoms. A water molecule is polar; therefore, water is a good solvent for other polar molecules. When the charged ends of water molecules interact with the charged groups of polar molecules, these polar molecules disperse in water.

Water is not a good solvent for nonpolar molecules, such as fats. A fat has no polar groups to interact with water molecules. An **emulsifier**, however, can cause a fat to disperse in water. An emulsifier contains molecules with both polar and nonpolar ends. When the nonpolar ends interact with the fat and the polar ends interact with the water molecules, the fat disperses in water, and an emulsion results (Fig. 4.6).

Bile salts (emulsifiers found in bile produced by the liver) are used in the digestive tract. Commercially produced emulsifiers include the wetting agent Tween and detergents. Today milk, such as 1% milk, has been homogenized so that fat droplets do not congregate and rise to the top of the container. Phospholipids are natural emulsifiers that are added to milk.

& Charged particle when interacted w/ Lipids

Figure 4.6 Emulsification.
An emulsifier contains molecules with both a polar and a nonpolar end. The nonpolar ends are attracted to the nonpolar fat, and the polar ends are attracted to the water. This causes droplets of fat molecules to disperse.

polar end

nonpolar end

emulsifier   fat

+

emulsion

*To test an unknown for fats and oils, use this procedure. Instead of water, use the unknown. If fats and oils are present, an oily spot appears.

Label three clean test tubes 1 to 3. Using the appropriate graduated transfer pipets add solutions to the test tubes as follows:

Tube 1
1. Add 3 ml of water and 1 ml of vegetable oil. Shake.
2. Observe for the initial dispersal of oil, followed by rapid separation into two layers.

   Is vegetable oil soluble in water? _____

3. Let the tube settle for 5 minutes. Label a microscope slide as 1.
4. Use a dropper to remove a sample of the solution that is just below the layer of oil. Place the drop on the slide, add a coverslip, and examine with the low power of your compound light microscope.
5. Record your observations in Table 4.8.

Tube 2
1. Add 2 ml of water, 3 ml of vegetable oil, and 1 ml of the available emulsifier. (Tween or bile salt solution) Shake.
2. Describe how the distribution of oil in tube 2 compares with the distribution in tube 1:

   Tube 2 emulsifier / oil stayed mixed, H_2O separated, after shaking   *Tube 1 → oil remained separated.*

3. Let the tube settle for 5 minutes. Label a microscope slide as 2.
4. Use a different dropper to remove a sample of the solution that is just below the layer of oil. Place the drop on the slide, add a coverslip, and examine with the low power of your compound light microscope.
5. Record your observations of the slides in Table 4.8.

Tube 3
1. Add 1 ml of milk and 2 ml of water. Shake well. Label a microscope slide as 3.
2. Use a different dropper to remove a sample of solution. Place a drop on a slide, add a coverslip, and examine with the lower power of your compound light microscope.
3. Record your observations in Table 4.8.

Table 4.8 Emulsification

Tube	Contents	Observations of Slides	Conclusions
1	Oil Water		
2	Oil Water Emulsifier	*emulsifier / oil stayed mixed but H_2O separated*	
3	Milk Water		

Conclusion: Emulsification of Lipids

- From your observations, conclude why the contents of each tube appear as they do under the microscope. Record your conclusions in Table 4.8.

- Explain the correlation between your macroscopic observations (how the tubes look to your unaided eye) and your microscopic observations. _____

_____

4.4 Testing Foods and Unknowns

It is common for us to associate the term *organic* with the foods we eat, including carbohydrate foods (Fig. 4.7), lipid foods (Fig. 4.8), and protein foods (Fig. 4.9). Though we may recognize foods as being organic, often we are not aware of what specific types of compounds are found in what we eat. In the following Experimental Procedure, you will use the same tests you used previously to determine the composition of everyday foods (unknowns).

Figure 4.7   Carbohydrate foods.

Figure 4.8   Lipid foods.

Figure 4.9   Protein foods.

Experimental Procedure: Testing Foods and Unknowns

Your instructor will provide you with several everyday foods including unknowns and your task is to:

1. Develop instructions for a procedure that will allow you to test the foods for carbohydrates (pages 37 and 39), proteins (page 42), and lipids (page 44) using the tests from this laboratory manual.

_____

_____

_____

_____

2. Have your instructor okay your procedure, and then conduct the necessary tests.
3. Record your results as positive or negative in Table 4.9.

Conclusions: Testing Foods and Unknowns

- Did any food test positive for only one of the organic compounds? _____

  Explain. _____

- What types of foods would you expect to test positive for more than one of the organic

  compounds studied in this laboratory? _____

- What type of carbohydrate might be found in an unknown that tests positive for the iodine

  test but negative for the Benedict's test? _____

  _____

Table 4.9   Testing Foods and Unknowns

Sample Name		Starch (Iodine)	Sugar (Benedict's)	Protein (Biuret)	Lipid (Brown Paper)
	Results:				
	Conclusion:				
	Results:				
	Conclusion:				

Application for Daily Living

Nutrition Labels

For good health, the diet should be balanced and should not contain an overabundance of either protein, carbohydrate, or fat. Everyone should become familiar with reading nutrition labels. A portion of one is shown in Figure 4.10.

It is very important to note the serving size and servings per container. For comparison purposes, you need to compare the same serving sizes. The % daily values are based on an intake of 2,000 calories per day. A calorie is an indication of the amount of energy provided, and the number of calories taken in should match the number required per day, unless you wish to gain weight.

Most people are extremely interested in the amount of fat provided by a food. One serving of this food provides 18% of the total fat required for the day. If you find this excessive, you could look for another macaroni and cheese that supplies less total fat and less saturated fat. If you decide that you prefer this product, you might wish to balance it with a food that contains very

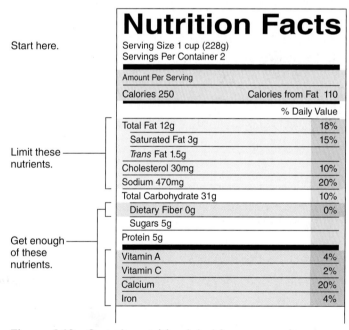

Figure 4.10   Sample nutrition label for macaroni and cheese.
As of January 2006, foods must include the quantity of trans fat. There is no % Daily Value for trans fats, and they should be avoided. See page 43.

little fat and one that provides a carbohydrate that contains fiber. This food contains no fiber. Complex carbohydrates containing fiber, such as those found in whole grains, are recommended and not those that contain simply starch.

Making wise decisions about the foods we buy can lead to a longer, healthier life.

Peptid Bond

enzyme

_R_

Starch

Glucose

Maltose

Polysacclides

Insulation, energy

glycerol

fats

DNA RNA

Protien

Immediatl

Benidicts

1. What type of bond joins amino acids to make peptides?

2. What type of protein speeds chemical reactions?

3. What group is different between types of amino acids?

4. If iodine solution turns blue-black, what substance is present?

5. If Benedict's reagent turns red, what substance is present?

6. When you digest food, starch is broken down to what molecule?

7. Is starch a monosaccharide or a polysaccharide?

8. What is the function of fat in humans?

9. What molecules are released when fat undergoes digestion?

10. What type of organic molecule is not soluble in water?

11. Name two organic molecules that are acids.

12. If Biuret reagent turns purple, what substance is present?

13. A student adds iodine solution to egg white and waits for a color change. How long will the student have to wait?

14. To test whether a sample contains glucose, what test should be done?

Thought Questions

15. Why is it necessary to shake a bottle of salad dressing before adding it to a salad?

16. An unknown sample is tested with both Biuret reagent and Benedict's reagent. Both tests result in a blue color. What has been learned? Why are these called negative results?

17. Starch and water are mixed together as ingredients for making gravy. Why doesn't starch automatically react with water to produce monosaccharides?

McGraw-Hill Access Science Website

An online encyclopedia of science and technology that provides information, including videos, that can enhance the laboratory experience.

www.accessscience.com

The following LearnSmart Labs contain exercises that are related to the content of this chapter:

• Chemical Composition of Cells

Find even more resources to learn the chapter concepts including animations, tutorial videos, and interactive practice questions.

5

Cell Structure and Function

Learning Outcomes

5.1 Anatomy of a Human Cell
- Using a model or drawing, identify the parts of a human cell and state a function for each part.

Prelab Question: Why is it beneficial for different parts of the cell to have specific functions?

5.2 Diffusion of Solutes
- Describe the process of diffusion as a physical phenomenon independent of the plasma membrane.
- Predict which solutes can cross the plasma membrane by diffusion and which cannot cross the plasma membrane by diffusion.

Prelab Question: Why is it beneficial for cellular substances to cross the plasma membrane by diffusion?

5.3 Osmosis: Diffusion of Water Across the Plasma Membrane
- Define osmosis and explain the movement of water across a membrane.
- Define isotonic, hypertonic, and hypotonic solutions and give examples in terms of NaCl concentrations.
- Predict the effect that solutions of different tonicities have on red blood cells.

Prelab Question: Why is 0.9% the usual tonicity of intravenous solutions?

5.4 Enzyme Activity
- Explain how enzymes have the ability to speed chemical reactions in cells.
- List some factors that can affect the speed of enzymatic reactions.

Prelab Question: Why is a warm body temperature advantageous to metabolism?

Application for Daily Living: **Dehydration and Water Intoxication**

Introduction

> **Planning Ahead** To save time, your instructor may have you start a boiling water bath (page 55) and the potato strip experiment (page 56) at the beginning of the laboratory.

We are accustomed to observing the outward appearance of human beings and rarely have an opportunity to become aware that humans are composed of small entities called **cells**. Though we often think that the heart, liver, or intestines enable the human body to function, it is actually cells that do the work of these organs. In Laboratory 3, you observed human cheek cells (see Figure 3.8) using a compound light microscope. The much more powerful **electron microscope** led to the discovery that cells are actually very complicated because they contain many **organelles** that carry out enzymatic functions.

The model of an animal cell available in the laboratory is based on electron micrographs. In today's laboratory, we review the structure and function of **organelles**, subcompartments of a cell, before actually observing how its outer surface, the **plasma membrane**, serves as a selective regulator of what enters and exits cells. We will discover that the passage of water into a cell depends on the difference in concentration of solutes (particles) between the **cytoplasm** (contents of a cell) and the surrounding medium or solution.

Enzymes are proteins that carry out metabolic reactions within organelles, and we will have an opportunity to marvel at the rapidity of an enzymatic reaction, even at room temperature.

5.1 Anatomy of a Human Cell

Figure 5.1 shows that an animal cell is partitioned into a number of compartments. Just like a house works more efficiently when each room has a specialized function, so does a cell with different compartments for varied functions.

Figure 5.1   Human (animal) cell.

Plasma membrane: outer surface that regulates entrance and exit of molecules
   protein
   phospholipid

CYTOSKELETON: maintains cell shape and assists movement of cell parts:
   Microtubules: cylinders of protein molecules present in cytoplasm, centrioles, cilia, and flagella
   Intermediate filaments: protein fibers that provide support and strength
   Actin filaments: protein fibers that play a role in movement of cell and organelles
   Centrioles: short cylinders of microtubules

Centrosome: microtubule organizing center that contains a pair of centrioles
   Vesicle: membrane-bounded sac that stores and transports substances
   Lysosome: vesicle that digests macromolecules and even cell parts
   Cytoplasm: semifluid matrix outside nucleus that contains organelles

NUCLEUS:
   Nuclear envelope: double membrane with nuclear pores that encloses nucleus
   Chromatin: diffuse threads containing DNA and protein
   Nucleolus: region that produces subunits of ribosomes

ENDOPLASMIC RETICULUM:
   Rough ER: studded with ribosomes, processes proteins
   Smooth ER: lacks ribosomes, synthesizes lipid molecules
   Ribosomes: particles that carry out protein synthesis

Mitochondrion: organelle that carries out cellular respiration, producing ATP molecules

Polyribosome: string of ribosomes simultaneously synthesizing same protein

Golgi apparatus: processes, packages, and secretes modified cell products

Learning the Organelles of the Cell

Identify these parts of a cell:

Composition and Function

Structure

1. Stack of membranous saccules; functions in processing, packaging, and distribution of molecules

   Golgi Apparatus

2. Membranous sacs; storage and transport of substances

   Vesicles

3. Have a double membrane; responsible for cellular respiration and production of ATP molecules

   mitochondria

4. Particles that carry out protein synthesis

   Ribosomes

5. Outer surface that regulates entrance and exit of molecules

   plasma membrane

6. Region in nucleus that produces subunits of ribosomes

   Nucleolus

7. Short cylinders, present in centrosomes, of unknown function

   Centrioles

8. Central body, having diffuse threads of DNA and protein

   Chromatin

9. Vesicle that digests macromolecules and even cell parts

   Lysosomes

10. Composed of microtubules, actin filaments, and intermediate filaments; responsible for the shape of the cell and movement of its parts

    Cytoskeleton

11. Membranous saccules and canals having no ribosomes; synthesizes lipid molecules

    Smooth E.R.

Learning That the Organelles Work Together

1. Imagine that this cell produces digestive enzymes that are sent to the digestive tract:

   a. Which part of the endoplasmic reticulum would produce these enzymes? _Rough_

   b. How would they be transported to another part of the cell? _Vesicles_

   c. Which organelle would process and package these enzymes for export? _golgi Apparatus_

2. Imagine that this cell produces a sex hormone (a lipid molecule):

   a. Which part of the endoplasmic reticulum would produce these lipid molecules? _Smooth_

   b. How would they be transported to another part of the cell? _Rough_

   c. Which organelle would process and package this hormone for export? _golgi Apparatus_

3. The nucleus produces the subunits of ribosomes.

   a. Where in the nucleus are the subunits produced? _Nucleolus_

   b. What part of the nuclear envelope allows them to get out of the nucleus? _Nuclear pore_

   c. Where do the subunits go and what happens to them? _____

4. How a cell breaks down engulfed substances.

   a. *Label Figure 5.1* where a vesicle is forming in order to take in a substance that will be digested.

   b. This vesicle will fuse with a _Lysosome_ that contains digestive enzymes.

5.2 Diffusion of Solutes

Diffusion is the random movement of molecules from the area of higher concentration to the area of lower concentration until they are equally distributed. A **solution** contains a molecule called the solute dissolved in a solvent (a liquid, most often water).

Environmental Factors and Diffusion of Molecules

If you spray a deodorant in one corner of the room, it will soon spread to fill the room because diffusion has occurred. (Notice, therefore, that diffusion can occur independently of a plasma membrane.) Environmental factors such as temperature and the resistance of the medium can affect the speed of diffusion. Air offers little resistance to random motion of molecules, followed by a liquid, and then by any type of a solid.

> *Observation: Environmental Factors and Diffusion*

You will calculate the speed of diffusion (1) through a semisolid gel and (2) through a liquid. Hypothesize whether you expect diffusion to occur faster through a semisolid or through a liquid, and give a reason for

your hypothesis. _____

_____

Diffusion Through a Semisolid

1. Observe a petri dish containing 1.5% gelatin (or agar) to which a crystal of potassium permanganate ($KMnO_4$) was added previously (time zero) in the center depression.
2. Obtain time zero from your instructor and record time zero and the final time (now) in Table 5.1. Calculate

   the length of time in hours and minutes; then convert the entire time to hours. _____ hr
3. Using a ruler placed over the petri dish, measure (in mm) the movement of color from the

   center of the depression outward in one direction. _____ mm

4. Calculate the speed of diffusion. _____ mm/hr
   (Divide the number of millimeters by the number of hours.)
5. Record all data in Table 5.1.

> ⚠️ **Potassium permanganate KMnO₄**
> $KMnO_4$ is highly poisonous and is a strong oxidizer. Avoid contact with skin and eyes and wash combustible materials. If spillage occurs, wash all surfaces thoroughly, $KMnO_4$ will also stain clothing.

Diffusion Through a Liquid

1. Add water to a glass petri dish.
2. Place the petri dish over a thin, flat ruler.
3. With tweezers, add a crystal of potassium permanganate ($KMnO_4$) directly over a millimeter measurement line. Note time zero in Table 5.1.
4. After 10 minutes, note the distance the color has moved (Fig. 5.2). Record the final time, length of time, and distance moved in Table 5.1.
5. Calculate the speed of diffusion by multiplying the length of time and the distance moved by 6.

   _____ mm/hr. Record in Table 5.1.   *measure in cm how much from edge*

Figure 5.2 Process of diffusion.
Diffusion is apparent when dye molecules have equally dispersed.

a. Crystal of dye in a semisolid.   b. Dye molecules diffuse.   c. Dye molecules are evenly distributed.

Table 5.1 Speed of Diffusion

Medium	Time Zero	Final Time	Length of Time (hr)	Distance Moved (mm)	Speed of Diffusion (mm/hr)
Semisolid	2mm	5min 4mm	10min Same		
Liquid	4mm	2Cm	3Cm		

Conclusions: Diffusion

- Why did the dye molecules move rather than stay where they were originally? _____

_____

- In which experiment was diffusion the fastest? _____

- What accounts for the difference in speed? _____

_____

The Plasma Membrane and Diffusion of Molecules

The plasma membrane regulates the passage of molecules into and out of cells. It is said to be **selectively permeable,** because only small, noncharged molecules can diffuse across the plasma membrane without assistance (Fig. 5.3a). Carriers, proteins embedded in plasma membrane, can assist the passage of molecules across a membrane. Each carrier is specific to a particular molecule. During facilitated transport, a carrier (or a channel protein) assists a molecule diffusing across the membrane (Fig. 5.3b). During active transport, a carrier assists a molecule moving the opposite to diffusion: from lower concentration outside the cell to higher concentration inside the cell and energy is required (Fig. 5.3c). In other words, energy is not required when a molecule is diffusing down its concentration gradient and, otherwise, energy is required.

Figure 5.3 Passage of molecules across a plasma membrane.
a. During diffusion, molecules move from the higher to the lower concentration. **b.** During facilitated transport, carrier proteins transport molecules from the higher to the lower concentration. **c.** During active transport, molecules move from the lower to the higher concentration; a protein carrier and energy are required.

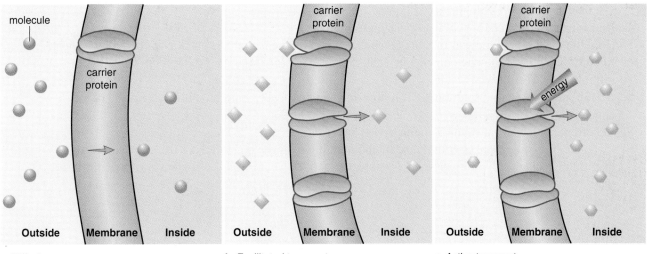

a. Diffusion   b. Facilitated transport   c. Active transport

In this experiment, an artificial membrane is used to simulate the plasma membrane; the artificial membrane is also semipermeable. Only certain molecules can cross the membrane. The artificial membrane contains no carriers to assist the movement of molecules across the membrane.

Notice in Figure 5.4 that at the start of this experiment, glucose (small molecule) and starch (large molecule) will be inside the membranous bag and iodine (small molecule) will be outside the bag. Hypothesize which molecules will cross the membrane and in which direction they will move.

to inside the bag ___iodine will enter bag, H₂O___

to outside the bag ___Nothing will exit bag___

Figure 5.4 Diffusion experiment.

rubber band

open end of bag

membranous bag

iodine solution

glucose and starch

closed end of bag

Diffusion Through a Selectively Permeable Membrane

At the start of this experiment:

1. Cut a piece of membranous tubing approximately 40 cm (approximately 16 in) long. Soak the tubing in water until it is soft and pliable.
2. Close one end of the tubing with two knots.
3. Fill the bag halfway with glucose solution.
4. Add four full droppers of starch suspension to the bag.
5. Hold the open end while you mix the contents of the bag. Rinse off the outside of the bag with water.
6. Record the color of the bag contents in Table 5.2.
7. Fill a beaker 2/3 full with water.
8. Add droppers of iodine solution (IKI) to the water in the beaker until an amber (tealike) color is apparent.
9. Record the color of the solution in the beaker in Table 5.2.
10. Place the bag in the beaker with the open end hanging over the edge. Secure the open end of the bag to the beaker with a rubber band as shown (Fig. 5.4). Make sure the contents do not spill into the beaker.

After about 5 minutes, at the end of the experiment:

11. You will note a color change. Record the present color of the bag contents and the beaker contents in Table 5.2.
12. Obtain a small test tube. Using a graduated transfer pipet, draw 1 ml from the bottom of the beaker (near the bag) and place it in the test tube. Using a designated transfer pipet, add 3 ml of Benedict's reagent. Heat in a boiling water bath for 5 to 10 minutes, observe any color change, and record your results as positive or negative in Table 5.2. (Optional use of glucose test strip: Dip glucose test strip into beaker. Compare stick with chart provided by instructor.)
13. Remove the dialysis bag from the beaker. Dispose of it and the used Benedict's reagent solution in the manner directed by your instructor.

> ⚠ **Benedict's reagent** Benedict's reagent is highly corrosive. Exercise care in using this chemical. If any should spill on your skin, wash the area with mild soap and water. Follow your instructor's directions for disposal of this chemical.

Table 5.2 Solute Diffusion Through a Membrane

	At Start of Experiment		At End of Experiment		
	Contents	Color	Color	Benedict's Test (+) or (−)	Conclusion
Bag	Glucose Starch	*Clear*		———	
Beaker	Water Iodine	*Tea Color*	*Light Blue Reagent*	*⊖*	*Neg. for glucose*

Conclusions: Plasma Membrane and Diffusion of Solutes

- Based on the color change noted in the bag, conclude what molecule diffused across the membrane from the beaker to inside the bag, and record your conclusion in Table 5.2.
- From the results of the Benedict's test on the beaker contents, conclude what molecule diffused across the membrane from the bag to the beaker, and record your conclusion in Table 5.2.
- Which molecule did not diffuse across the membrane from the bag to the beaker? _____

  Explain. _____

5.3 Osmosis: Diffusion of Water Across the Plasma Membrane

Osmosis is the diffusion of water molecules across a selectively permeable membrane. Like other molecules, water follows its concentration gradient and moves from a region of higher concentration to a region of lower concentration. A new finding has been that water passes through a channel protein called an **aquaporin** because of its specificity. Therefore, osmosis (the diffusion of water) is a form of facilitated transport.

Tonicity is the relative concentration of solute (e.g., salt molecules) and solvent (water molecules) outside the cell compared to inside the cell.[1] An **isotonic solution** has the same concentration of solute (and therefore of water) as the cell. When cells are placed in an isotonic solution, there is no net movement of water. A **hypertonic solution** has a higher solute (therefore, lower water) concentration than the cell. When cells are placed in a hypertonic solution, water moves out of the cell into the solution. A **hypotonic solution** has a lower solute (therefore, higher water) concentration than the cell. When cells are placed in a hypotonic solution, water moves from the solution into the cell.

The next two Experimental Procedures explore tonicity using potato strips and red blood cells.

[1]Percent solutions are grams of solute per 100 ml of solvent. Therefore, a 10% solution is 10 g of sugar with water added to make up 100 ml of solution.

Experimental Procedure: Tonicity and Potato Strips

This procedure runs for 1 hour. Prior setup can maximize your time efficiency.

1. Cut two strips of potato, each about 7 cm long and 1 1/2 cm wide.
2. Label two test tubes 1 and 2. Place one potato strip in each tube.
3. Fill tube 1 with water to cover the potato strip.
4. Fill tube 2 with 10% sodium chloride (NaCl) to cover the potato strip. NaCl is table salt.
5. After 1 hour, observe each strip for limpness (water loss) or stiffness (water gain). Which tube

   has the limp potato strip? ___2___ Why did water diffuse out of the potato strip? _____

   Which tube has the stiff potato strip? ___1___ Why did water diffuse into the potato strip?

   #1 - Crisp - potato absorbing water

   #2 - potato releasing water

Red Blood Cells (Animal Cells)

A solution of 0.9% NaCl is isotonic to red blood cells. In such a solution, red blood cells maintain their normal appearance (Fig. 5.5*a*). A solution greater than 0.9% NaCl is hypertonic to red blood cells. In such a solution, the cells shrivel up, a process called **crenation** (Fig. 5.5*b*). A solution of less than 0.9% NaCl is hypotonic to red blood cells. In such a solution, the cells swell to bursting, a process called **hemolysis** (Fig. 5.5*c*).

Complete Table 5.3 by following these instructions. In the second column, state whether the solution is isotonic, hypertonic, or hypotonic to red blood cells. In the third column, hypothesize the effect on the shape of the cell after being in this solution. In the fourth column, explain why you hypothesized this outcome. Base your explanation on the movement of water.

Table 5.3 Effect of Tonicity on Red Blood Cells

Concentration (NaCl)	Tonicity	Effect on Cells	Explanation
0.9%	isotonic	No effect	Same amount of H₂O moves in as what moves out
Higher than 0.9%	hypertonic	cells shrivel	water leaves cell
Lower than 0.9%	hypotonic	cell swells / bursts	water enters cell

Figure 5.5 Tonicity and red blood cells.

| 15,000× | 15,000× | 15,000× |

a. Isotonic solution.
Red blood cell has normal appearance due to no net gain or loss of water.

b. Hypertonic solution.
Red blood cell shrivels due to loss of water.

c. Hypotonic solution.
Red blood cell fills to bursting due to gain of water.

Three numbered and stoppered test tubes are on display. Each test tube contains NaCl and a few drops of blood.

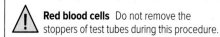

Red blood cells Do not remove the stoppers of test tubes during this procedure.

1. Shake each tube as shown in Figure 5.6*a*. Then place the tube in front of your lab manual as shown in Figure 5.6*b*. Determine whether you can see the print on the page and record your decision under Print Visibility in Table 5.4.
2. Dependent on how difficult it is to see the print, which tube is hypotonic (less than 0.9% NaCl), hypertonic (10% NaCl), or isotonic (0.9% NaCl)? Record your deduction under Tonicity in Table 5.4.
3. In the last column of Table 5.4, explain how you arrived at this deduction.

Table 5.4 Print Visibility and Tonicity

Tube	Print Visibility	Tonicity	Explanation
.9% 1	NO VISIbility	isotonic	
10% 2	NO VISIbility	hyper	
H2O 3	VISIble	hypo	

Figure 5.6 Proof of hemolysis.
a. Shake the tube as shown here. **b.** Once the red blood cells burst, you can read print placed behind a tube of diluted blood.

stopper

a. Gently invert tube several times.

b. Determine whether print is readable through tube.

5.4 Enzyme Activity H_2O / peroxide

Enzymes are organic catalysts that speed metabolic reactions, either degradation or synthesis (Fig. 5.7). Each enzyme has a three-dimensional shape that accommodates its substrate(s), the reactant(s) in the enzyme's reaction. This shape, therefore, determines the specificity of the enzyme and is important to the action of the enzyme. Although the shape of the enzyme and its substrate are compatible, the favored model for enzyme action suggests that the enzyme initially interacts with its substrate, changes shape slightly to improve the interaction, and then proceeds with a more efficient reaction. Certain environmental effects ensure that enzymes can function speedily. A warm temperature, sufficient enzyme and substrate concentrations, and the correct pH (whether the solution should be acidic, basic, or neutral) are all important. Each enzyme has a pH at which the speed of the reaction is optimum. Any pH higher or lower than the optimum affects the shape of the enzyme, leading to reduced activity.

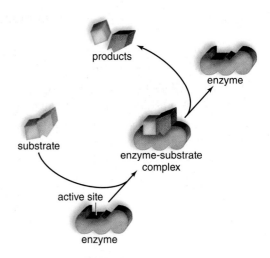

Figure 5.7 Enzymatic action.
During degradation, the substrate is broken down into smaller products.

Experiment with the Enzyme Catalase

In the Experimental Procedure that follows, you will be working with the enzyme catalase. **Catalase** is present in cells, where it speeds the breakdown of the toxic chemical hydrogen peroxide (H_2O_2) to water and oxygen:

$$2H_2O_2 \xrightarrow{\text{catalase}} 2H_2O + O_2$$

hydrogen peroxide → water oxygen

In this example, the enzyme is catalase; the substrate that fits within the active site of the enzyme is hydrogen peroxide; and the products are water and oxygen. Catalase performs a useful function in organisms because hydrogen peroxide is harmful to cells. Hydrogen peroxide is a powerful oxidizer that can attack and denature cellular molecules such as DNA. Knowing its harmful nature, humans use hydrogen peroxide as a commercial antiseptic to kill germs. In reduced concentration, hydrogen peroxide is a whitening agent used to bleach hair and teeth. It is also used industrially to clean most anything from tubs to sewage.

Experimental Procedure: Catalase Activity

This Experimental Procedure tests the effects of pH on the activity of catalase. Potato will be the source of catalase. As the reaction occurs, easily observable bubbling will develop. Label two clean test tubes and use the appropriate graduated transfer pipets to follow these directions.

> ⚠️ **Protective eyewear** Protective eyewear should be worn for this procedure. HCl is a strong caustic acid. Exercise care in using this chemical, and follow your instructor's directions for disposal of this tube. If any acid should spill on your skin, rinse immediately with water.

Tube 1
1. Add 2 ml of distilled water to Tube 1. (This is **neutral pH.**)
2. Macerate a cube (1 cc) of potato using a mortar and pestle. Transfer to the tube. *chop up*
3. Wait 3 minutes.
4. Add 4 ml of hydrogen peroxide. Record bubbling in Table 5.5 using 0 (no bubbling) or + signs (e.g., +, ++, +++, most bubbling).

Tube 2
1. Carefully add 2 ml of hydrochloric acid (5M HCl) to this tube.
2. Macerate a cube (1 cc) of potato using a mortar and pestle. Transfer to the tube.
3. Wait 3 minutes.
4. Add 4 ml hydrogen peroxide. Record the degree of bubbling in Table 5.5.

Table 5.5 Effect of pH on Catalase Activity

Tube	Contents	Bubbling*	Explanation
1 *water has no effect*	Distilled water Potato, macerated Hydrogen peroxide	*Lots of bubbling*	*Reaction of catalase is evident after adding peroxide*
2	Hydrochloric acid Potato, macerated Hydrogen peroxide	*Not as much Bubbling*	*Changed shape of catalase. decreases activity w/ peroxide*

*Ignore large bubbles and look for small bubbles.

Denaturation - changes Shape

Conclusions: Catalase Activity

- Give explanations for your results in Table 5.5.
- From your test results, decide whether enzymes can be negatively affected by environmental conditions and explain your answer. _____

_____

Application for Daily Living

Dehydration and Water Intoxication

Most people have heard of dehydration, but they may not realize that dehydration occurs because blood has become hypertonic to cells and, in response, the cells lose water. Dehydration can be due to excessive sweating, perhaps during exercise, or it can be a side effect of many illnesses that cause prolonged vomiting or diarrhea. The signs of moderate dehydration are a dry mouth, sunken eyes, and skin that will not bounce back after light pinching. Most people have never heard of water intoxication, which occurs when blood becomes hypotonic to cells. In response to hypotonicity, cells gain water. Water intoxication can lead to pulmonary edema (the lungs gain water) and swelling in the brain. In extreme cases, it is fatal. It can be due to an intake of too much pure water during vigorous exercise, such as a marathon race. The cure is introducing an intravenous solution containing high amounts of sodium in a hospital setting. To prevent both dehydration and water intoxication, athletes should replace lost fluids slowly. Pure water is a good choice if the exercise period is short. Low-sodium solutions, such as sports drinks, are a good choice for longer-duration events like marathons.

Laboratory Review 5

PROCESS PROTEIN **1.** What is the function of rough endoplasmic reticulum?

lysosome **2.** Which organelle carries on intracellular digestion?

Brain of cell **3.** What is the function of the nucleus?

Ribosome **4.** Which organelle is responsible for protein synthesis?

Diffusion **5.** What term is used to describe the movement of molecules from an area of higher concentration to an area of lower concentration?

osmosis **6.** What is the name for the movement of water across a selectively permeable membrane?

hypertonic **7.** Is 10% NaCl isotonic, hypertonic, or hypotonic to red blood cells?

Swelling / swell **8.** What appearance will red blood cells have when they are placed in 0.0009% NaCl?

out of cell **9.** In which direction does water move when cells are placed in a hypertonic solution?

Substrates **10.** The active site of an enzyme brings together the _____ of a reaction.

Reduced activity / slows **11.** In general, what does unfavorable environmental conditions due to the speed of an enzymatic reaction?

Shape **12.** An unfavorable pH causes an enzyme to lose its normal _____.

Thought Questions

13. If a dialysis bag filled with water is placed in a starch solution, what do you predict will happen to the weight of the bag over time? Why?

14. Ocean water is hypertonic to the internal environment of the body. Predict what would happen to your cells if you consumed large quantities of ocean (salt) water.

6

Body Tissues

Learning Outcomes

6.1 Epithelial Tissue
- Identify slides and models or diagrams of various types of epithelium.
- Tell where a particular type of epithelium is located in the body, and state a function.

Prelab Question: Where in the digestive tract would you find epithelial tissue? What various functions could it have along the tract?

6.2 Connective Tissue
- Identify slides and models or diagrams of various types of connective tissue.
- Tell where a particular connective tissue is located in the body, and state a function.

Prelab Question: Name six types of connective tissue and give a location for each type in the body.

6.3 Muscular Tissue
- Identify slides and models or diagrams of three types of muscular tissue.
- Tell where each type of muscular tissue is located in the body, and state a function.

Prelab Question: Associate each type of muscular tissue with a particular organ (or organs) in the body.

6.4 Nervous Tissue
- Identify a slide and model or diagram of a neuron.
- Tell where nervous tissue is located in the body, and state a function.

Prelab Question: Which part of a neuron do you expect to see concentrated in the brain?

6.5 Tissues Form Organs
- Use a portion of the digestive tract to show that organs can contain four different types of tissue: epithelial, connective, muscular, and nervous tissues.

Prelab Question: Give an example of the levels of organization in the body.

Application for Daily Living: **Tissue Engineering**

Introduction

In our last laboratory, we studied a generalized cell, one that has no particular specialization. The cells in our body are usually specialized in structure and function, and these specialized cells congregate together in a **tissue.** So, for example, muscular tissue contains muscle cells that all have the same appearance and perform the same task. That is, each muscle cell can contract, and because of this, muscular tissue can contract.

While some **organs** of the body are known for having a particular type of tissue—for example, the brain contains nervous tissue and a muscle contains muscular tissue—other organs have several types of tissues. We will end this laboratory by taking a look at a portion of the digestive tract, which has several types of tissues. This will help you realize that tissues are not isolated in the body; instead, they are a part of a specialized organ.

6.1 Epithelial Tissue

Epithelial tissue (epithelium) forms a continuous layer, or sheet, over the entire body surface and most of the body's inner cavities. Externally, it forms a covering that protects the animal from infection, injury, and drying out. Some epithelial tissues produce and release secretions. Others absorb nutrients.

The name of an epithelial tissue includes two descriptive terms: the shape of the cells and the number of layers. The three possible shapes are squamous, cuboidal, and columnar. With regard to layers, an epithelial tissue may be simple or stratified. **Simple** means that there is only one layer of cells; **stratified** means that cell layers are placed on top of each other. Some epithelial tissues are **pseudostratified,** meaning that they only appear to be layered. Epithelium may also have cellular extensions called **microvilli,** or hairlike extensions called **cilia.** In the latter case, "ciliated" may be part of the tissue's name.

Observation: Simple Squamous Epithelium

Simple squamous epithelium is a single layer of thin, flat, many-sided cells, each with a central nucleus. It lines internal cavities, the heart, and all the blood vessels. It also lines parts of the urinary, respiratory, and male reproductive tracts.

1. Study a model or diagram of simple squamous epithelium (Fig. 6.1). What does *squamous* mean?

   _____

2. Examine a prepared slide of squamous epithelium. Under low power, note the close packing of the flat cells. What shapes are the cells?

   _____

3. Under high power, examine an individual cell, and identify the plasma membrane, cytoplasm, and nucleus.

Observation: Skin and Stratified Epithelium

The skin covers the entire exterior of the human body, and because of its complex nature it is sometimes called the **integumentary system.** Skin functions include protection, water retention, sensory reception, body temperature regulation, and vitamin D synthesis.

Skin Model

Study a model or diagram and also a prepared slide of the skin. With the help of Figure 6.2, identify the two skin regions and the subcutaneous layer.

1. **Epidermis:** This region is composed of **stratified squamous epithelium.** The outer cells are nonliving and create a waterproof covering that prevents excessive water loss. These cells are always being replaced because an inner layer of the epidermis is composed of living cells that constantly produce new cells.

Figure 6.1 Simple squamous epithelium.
Simple squamous epithelium lines blood vessels and various tracts.

Simple squamous
• lining of lungs, blood vessels
• protects

250×

basement membrane

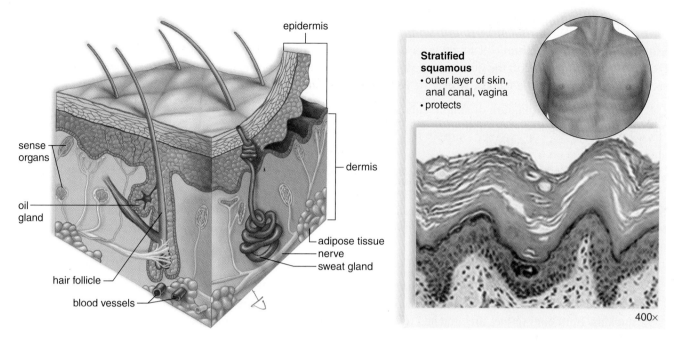

Figure 6.2 Skin.
Notice in the art and micrograph that the epidermis is composed of stratified squamous epithelium.

epidermis

dermis

adipose tissue
nerve
sweat gland

sense organs

oil gland

hair follicle

blood vessels

Stratified squamous
• outer layer of skin, anal canal, vagina
• protects

400×

2. **Dermis:** This region is a connective tissue (p. 68) containing blood vessels, nerves, sense organs, and the expanded portions of oil (sebaceous) and sweat glands and hair folicles. List the structures you can identify on your slide:

_____

_____

3. **Subcutaneous layer:** This is a layer of loose connective tissue and adipose tissue (p. 68) that lies beneath the skin proper and serves to insulate and protect inner body parts. This layer is not part of the skin.

Stratified Squamous Epithelium

As would be expected from its name, stratified squamous epithelium consists of many layers of cells. The innermost layer produces cells that are first cuboidal or columnar in shape, but as the cells push toward the surface, they become flattened.

As mentioned, the outer region of the skin is stratified squamous epithelium. As the cells move toward the surface, they flatten, begin to accumulate a protein called **keratin,** and eventually die. Keratin makes the outer layer of epidermis tough, protective, and able to repel water.

The linings of the mouth, throat, anal canal, and vagina are also stratified epithelium. The outermost layer of cells surrounding the cavity is simple squamous epithelium. In these organs, this layer of cells remains soft, moist, and alive.

1. Examine a slide of stratified squamous epithelium.

2. Approximately how many layers of cells do you see? _____

3. Which layers of cells seem to be squamous epithelium? _____

Observation: Simple Cuboidal Epithelium

Simple cuboidal epithelium is a single layer of cube-shaped cells, each with a central nucleus. It is found in tubules of the kidney and in the ducts of many glands, where it has a protective function. It also occurs in the secretory portions of some glands, where the tissue produces and releases secretions.

1. Study a model or diagram of simple cuboidal epithelium (Fig. 6.3).
2. Examine a prepared slide of simple cuboidal epithelium. Move the slide until you locate cube-shaped cells that line a lumen (cavity).

Observation: Simple Columnar Epithelium

Simple columnar epithelium is a single layer of tall, cylindrical cells, each with a nucleus near the base. This tissue, which lines the digestive tract from the stomach to the anus, protects, secretes, and allows absorption of nutrients.

1. Study a model or diagram of simple columnar epithelium (Fig. 6.4).
2. Examine a prepared slide of simple columnar epithelium. Find tall and narrow cells that line a lumen. Under high power, focus on an individual cell. Identify the plasma membrane, the cytoplasm, and the nucleus. Epithelial tissues are attached to underlying tissues by a basement membrane composed of extracellular material containing protein fibers.
3. The tissue you are observing contains mucus-secreting cells. Search among the columnar cells until you find a **goblet cell,** so named because of its goblet-shaped, clear interior. This region contains mucus, which may be stained a light blue. In the living animal, the mucus is discharged into the gut cavity and protects the lining from digestive enzymes.

Figure 6.3   Simple cuboidal epithelium.
Simple cuboidal epithelium lines kidney tubules and the ducts of many glands.

Simple cuboidal
• lining of kidney tubules, various glands
• absorbs molecules

lumen

250×

basement membrane

Figure 6.4   Simple columnar epithelium.
Simple columnar epithelium lines the digestive tract. Goblet cells among the columnar cells secrete mucus.

Simple columnar
• lining of small intestine, oviducts
• absorbs nutrients

basement membrane

250×

goblet cell secretes mucus

basement membrane

Pseudostratified ciliated columnar epithelium

appears to be layered, though actually all cells touch the basement membrane. Many cilia are located on the free end of each cell (Fig. 6.5). In the human trachea, the cilia wave back and forth, moving mucus and debris up toward the throat so that it cannot enter the lungs. Smoking destroys these cilia, but they will grow back if smoking is discontinued.

1. Study a model or diagram of pseudostratified ciliated columnar epithelium (Fig. 6.5).
2. Examine a prepared slide of pseudostratified ciliated columnar epithelium. Concentrate on the part of the slide that resembles the model. Identify the cilia.

Pseudostratified ciliated columnar
- lining of trachea
- sweeps impurites toward throat

250×

cilia

goblet cell secretes mucus

basement membrane

Figure 6.5 Pseudostratified ciliated columnar epithelium.
Pseudostratified ciliated columnar epithelium lines the trachea. The cilia help keep the lungs free of debris.

Summary

Identify the type of epithelial tissue.

a. _____

c. _____

b. _____

d. _____

6.2 Connective Tissue

Connective tissue joins different parts of the body together. There are four general classes of connective tissue: connective tissue proper (loose fibrous, dense fibrous, and adipose), bone, cartilage, and blood. All types of connective tissue consist of cells surrounded by a matrix that usually contains fibers. Elastic fibers are composed of a protein called elastin. Collagenous fibers contain the protein collagen.

Observation: Loose and Dense Fibrous Connective Tissue

There are several different types of connective tissue. We will study loose fibrous connective tissue, dense fibrous connective tissue, adipose tissue, bone, cartilage, and blood. **Loose fibrous connective tissue** supports *Cherry Blossoms* epithelium and many internal organs, such as muscles, blood vessels, and nerves (Fig. 6.6). Its presence allows organs to expand. **Dense fibrous connective tissue** contains many collagenous fibers packed together, as in tendons, which connect muscles to bones, and in ligaments, which connect bones to other bones at joints (Fig. 6.7).

1. Examine a slide of loose fibrous connective tissue, and compare it with Figure 6.7. What is the function of loose fibrous connective tissue? _____

    _____

2. Examine a slide of dense fibrous connective tissue, and compare it with Figure 6.8. What two types of structures in the body contain dense fibrous connective tissue?

    _____

    _____

Areolar — Loose connective

Figure 6.6 Loose fibrous connective tissue.
Loose fibrous connective tissue supports epithelium and many internal organs.

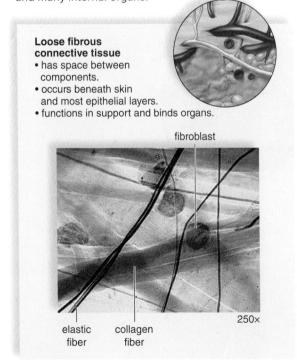

Loose fibrous
connective tissue
• has space between
  components.
• occurs beneath skin
  and most epithelial layers.
• functions in support and binds organs.

fibroblast

elastic   collagen   250×
fiber   fiber

Figure 6.7 Dense fibrous connective tissue.
Dense fibrous connective tissue is found in tendons and ligaments.

Dense fibrous
connective tissue
• has collagenous fibers
  closely packed.
• is in dermis of skin, tendons,
  and ligaments.
• functions in support.

collagen   nuclei of   400×
fibers   fibroblasts

Observation: Adipose Tissue

In **adipose tissue,** the cells have a large, central, fat-filled vacuole that causes the nucleus and cytoplasm to be at the perimeter of the cell (see Fig. 6.9). Adipose tissue occurs beneath the skin, where it insulates the body, and around internal organs, such as the kidneys and heart. It cushions and helps protect these organs.

1. Examine a prepared slide of adipose tissue. Why is the nucleus pushed to one side? _____

    _____

2. State a location for adipose tissue in the body. _____

    _____

    What are two functions of adipose tissue at this location? _____

    _____

Observation: Compact Bone

Compact bone is found in the bones that make up the skeleton. It consists of **osteons** (Haversian systems), with a **central canal,** and concentric rings of spaces called **lacunae,** connected by tiny crevices called **canaliculi.** The central canal contains a nerve and blood vessels, which service bone. The lacunae contain bone cells called **osteocytes,** whose processes extend into the canaliculi. Separating the lacunae is a matrix that is hard because it contains minerals, notably calcium salts. The matrix also contains collagenous fibers.

1. Study a model or diagram of compact bone (Fig. 6.8). Then look at a prepared slide and identify the central canal, lacunae, and canaliculi.

2. What is the function of the central canal and canaliculi? _____

    _____

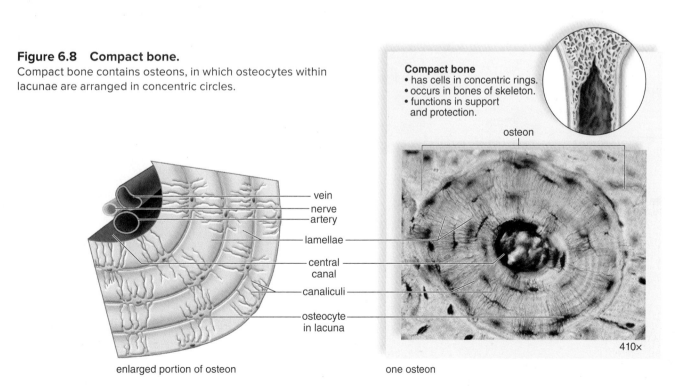

Figure 6.8   Compact bone.
Compact bone contains osteons, in which osteocytes within lacunae are arranged in concentric circles.

Compact bone
• has cells in concentric rings.
• occurs in bones of skeleton.
• functions in support and protection.

osteon

vein
nerve
artery
lamellae
central canal
canaliculi
osteocyte in lacuna

410×

enlarged portion of osteon

one osteon

Observation: Hyaline Cartilage

In **hyaline cartilage,** cells called **chondrocytes** are found in twos or threes in lacunae. The lacunae are separated by a flexible matrix containing weak collagenous fibers. Hyaline cartilage occurs in the nose and walls of respiratory passage; at ends of bone, including ribs. It functions in support and protection.

1. Study the diagram and photomicrograph of hyaline cartilage in Figure 6.9. Then study a prepared slide of hyaline cartilage, and identify the matrix, lacunae, and chondrocytes.
2. Compare slides of compact bone and hyaline cartilage. Which of these types of connective tissue is more

organized? _____

Why? _____

3. Which of these two types of connective tissue lends more support to body parts? _____

_____

Figure 6.9   Connective tissues associated with the knee.
The human knee provides examples of most types of connective tissue.

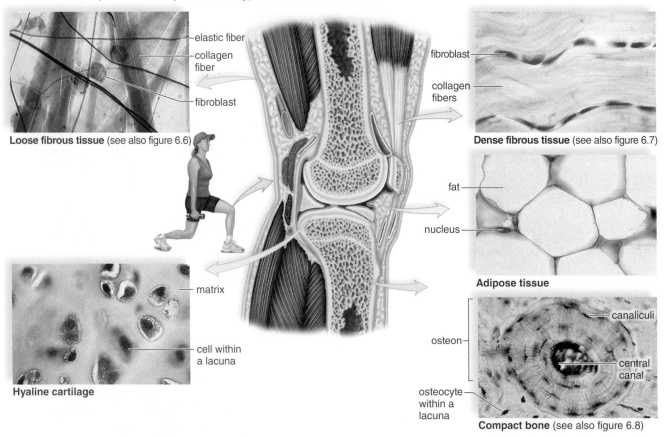

Loose fibrous tissue (see also figure 6.6)

elastic fiber
collagen fiber
fibroblast

fibroblast
collagen fibers

Dense fibrous tissue (see also figure 6.7)

fat
nucleus

Adipose tissue

matrix

cell within a lacuna

Hyaline cartilage

osteon
osteocyte within a lacuna

canaliculi
central canal

Compact bone (see also figure 6.8)

Blood is a connective tissue in which the matrix is an intercellular fluid called **plasma. Red blood cells** (erythrocytes) carry oxygen combined with the respiratory pigment hemoglobin. **White blood cells** (leukocytes) fight infection. Also present in blood are many small bodies, the **platelets,** which play a major role in clot formation.

1. Study a prepared slide of human blood. With the help of Figure 6.10, identify the numerous red blood cells and the less numerous but larger white blood cells, which appear faint because of the stain. As you scan your slide on high power, also look for the small platelets, the small objects scattered between the blood cells.

2. Try to identify a neutrophil, the most common type of white blood cell. A neutrophil has a multilobed nucleus.

Neutrophil   400×

Figure 6.10 Neutrophil.
A neutrophil is a spherical white blood cell about 10–14 μm in diameter that has a multilobed nucleus and fine, pink granules in the cytoplasm. It is capable of phagocytizing pathogens.

6.3 Muscular Tissue

Muscular (contractile) tissue is composed of cells called muscle fibers. Muscular tissue has the ability to contract, and contraction usually results in movement. The body contains skeletal, cardiac, and smooth muscle.

Skeletal muscle occurs in the muscles attached to the bones of the skeleton. The contraction of skeletal muscle is said to be **voluntary** because it is under conscious control. Skeletal muscle is striated; it contains light and dark bands. The striations are caused by the arrangement of contractile filaments (actin and myosin filaments) in muscle fibers. Each fiber contains many nuclei, all peripherally located.

Skeletal muscle
- has striated cells with multiple nuclei.
- occurs in muscles attached to skeleton.
- functions in voluntary movement of body.

1. Study a model or diagram of skeletal muscle (Fig. 6.11). Striations are present. You should see several muscle fibers, each marked with striations.

2. Examine a prepared slide of skeletal muscle. The striations may be difficult to make out, but bringing the slide in and out of focus may help.

striation   nucleus   400×

Figure 6.11 Skeletal muscle.
Skeletal muscle is striated and voluntary. Its cells are tubular and contain many nuclei.

Observation: Cardiac Muscle

Cardiac muscle is found only in the heart. It is called **involuntary** because its contraction does not require conscious effort. Cardiac muscle is striated in the same way as skeletal muscle. However, the fibers are branched and bound together at **intercalated disks,** where their folded plasma membranes touch. This arrangement aids communication between fibers.

1. Study a model or diagram of cardiac muscle (Fig. 6.12). Striations are present.
2. Examine a prepared slide of cardiac muscle. Find an intercalated disk. What is the function of

    cardiac muscle? _____

    _____

Observation: Smooth Muscle

Smooth muscle is sometimes called **visceral muscle** because it makes up the walls of the internal organs, such as the intestines and the blood vessels. Smooth muscle is involuntary because its contraction does not require conscious effort.

1. Study a model or diagram of smooth muscle (Fig. 6.13), and note the shape of the cells and the centrally placed nucleus. Smooth muscle has spindle-shaped cells. What does *spindle-shaped* mean?

    _____

    _____

2. Examine a prepared slide of smooth muscle. Distinguishing the boundaries between the different cells may require you to bring the slide in and out of focus.

Figure 6.12   Cardiac muscle.
Cardiac muscle is striated and involuntary. Its branched cells join at intercalated disks.

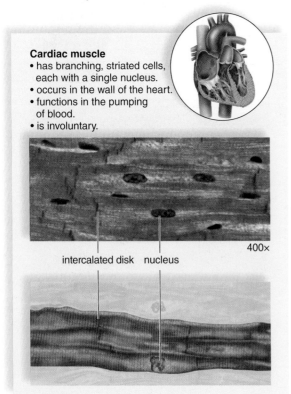

Figure 6.13   Smooth muscle.
Smooth muscle is nonstriated and involuntary. This type of muscle is composed of spindle-shaped cells.

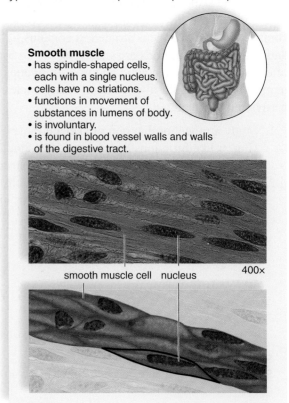

6.4 Nervous Tissue

Nervous tissue is found in the brain, spinal cord, and nerves. Nervous tissue is composed of two types of cells: **neurons** that transmit messages and **neuroglia** that largely service the neurons. Motor neurons, which take messages from the spinal cord to the muscles, are often used to exemplify typical neurons (Fig. 6.14). Motor neurons have several **dendrites,** processes that take signals to a **cell body,** where the nucleus is located, and an **axon** that takes nerve impulses away from the cell body. Long axons are called nerve fibers.

Observation: Nervous Tissue

1. Study a model or diagram of a neuron, and then examine a prepared slide. Depending on the magnification, you may not be able to see neuroglial cells because they are small and cannot be seen at low magnification.
2. Identify the the parts of a motor neuron in Figure 6.14 and also the neuroglia. *Label Figure 6.14a.*
3. Explain the appearance and function of the parts of a motor neuron:

    a. Dendrites _____

    b. Cell body _____

    c. Axon _____

Figure 6.14   Motor neuron anatomy.

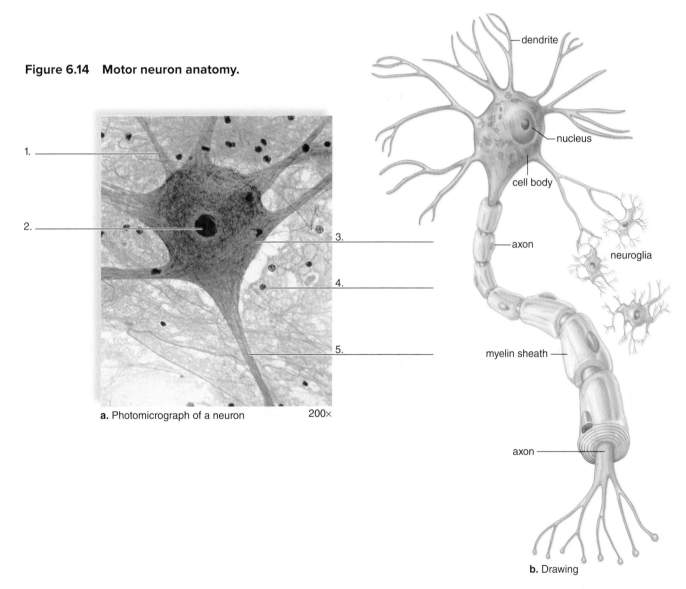

a. Photomicrograph of a neuron   200×

b. Drawing

6.5 Tissues Form Organs

There are only four types of tissues in the body: epithelial, connective, muscular, and nervous. To simplify matters, epithelial tissues form protective coverings; connective tissues connect other tissues; muscular tissues contract; and nervous tissues are sensitive to stimuli. None of these tissues occur in a vacuum. They are integral parts of an organ. We will take as our example a portion of the digestive tract shown in Figure 6.15.

The inner layer of the tract is called the mucosa. The mucosa is an inner lining of epithelial cells, some of which are glandular and secrete mucus. Epithelium performs a protective function; the mucus protects the other layers of the tract. Occasionally along the tract, some other epithelial cells are glandular and secrete digestive juices.

Beneath the epithelial lining, the submucosa is a connective tissue layer. Connective tissue binds together other tissues. As is often the case, lymphatic vessels, blood vessels, and nerves run along within the connective tissue layer.

Next, the muscularis has two layers of muscle tissue; the contraction of these layers mixes food and digestive juices in the tract. It also pushes the food along the tract. Muscles don't contract unless they are stimulated to do so by nerves; nerves occur between the two layers of muscle.

Finally, the serosa is connective tissue that supports the digestive organs within the abdominal cavity.

Figure 6.15 Wall of the gastrointestinal tract.
The wall of the gastrointestinal tract contains the four layers noted.

Mucosa
inner mucous membrane
layer that is modified
according to the
digestive organ

Submucosa
broad band of loose
connective tissue that
contains nerves, blood,
and lymphatic vessels

Muscularis
two layers of
smooth muscle

Serosa
thin, outermost epithelial
tissue overlying
connective tissue

nerve   lymphatic
supply   vessel

lumen

artery

vein

You can apply what you know about the tissues in the digestive tract to any other organ in the body. No matter where you find these tissues, what do you expect them to do?

Tissue: **General Function:**

Epithelial tissue   _____

Connective tissue   _____

Muscular tissue   _____

Nervous tissue   _____

In the next laboratory, we will study the various systems of the body and learn that each **organ system,** such as the digestive system, contains several different types of organs. The digestive system contains the esophagus, stomach, small intestine, and large intestine. Use the digestive system to illustrate the levels of organization of the human body from an organ system to cells:

Name of organ system: _____

Name of organs: _____

Types of tissues in each organ: _____

Type of cell in a particular tissue type: _____

Application for Daily Living

Tissue Engineering

Just a few years ago, scientists believed that transplant tissues and organs had to come directly from humans. Now, however, tissue engineering is demonstrating that it is possible to make replacement tissues and organs in the lab. For example, skin products have now been approved for use in humans. One is composed of dermal cells growing on a degradable polymer, which can be used to temporarily cover the wounds of burn patients while their own skin regenerates (Fig. 6.16). The other uses only live human skin cells to treat diabetic leg and foot ulcers. Similarly, the damaged cartilage of a knee can be replaced with a tissue produced after chondrocytes are harvested from a patient. Soon to come is a host of other products, including replacement corneas, heart valves, bladder valves, and breast tissue.

After nine years, researchers were able to produce a working urinary bladder in the laboratory. This bladder can be implanted in humans whose bladders have been damaged by accident or disease. Another group of scientists have been able to grow arterial blood vessels in the lab using a pig's small intestine as the mold. Tissue engineers recently announced they had produced a human heart in the lab.

Figure 6.16   Acid burn victim.
This young woman, whose face was damaged by acid, received skin, produced in the laboratory, as a transplant.

tissue   1. What is the name for a group of cells that has the same structural characteristics and performs the same functions?

simple squamous   2. Which type of epithelium has flattened cells?

Trachea ✴ 3. Name a body location for pseudostratified ciliated columnar epithelium.

Produces mucous/Protects
stratified squamous ✴4. What is the function of goblet cells? Produces mucous, Protects lining from digestive Enzymes

epithelial   5. What type of tissue occurs in the epidermis of the skin?

Rib cage - end of long bone   6. Name a body location for hyaline cartilage.

adipose   7. The cells of which tissue have a large, central, fat-filled vacuole?

cardiac ✴ 8. What type of muscular tissue is involuntary and striated?

Digestive Tract   9. Name a body location for smooth muscle.

Skeletal   10. What types of muscular tissue are striated?

Brain/spine   11. Name a body location for nervous tissue.

Cell body ✴ 12. Where is the nucleus located in a nerve cell?

Smooth   13. What type of tissue accounts for the movement of food along the digestive tract?

Dermis ✴14. Which skin layer contains blood vessels? connective

axon   15. What portion of a nerve cell transmits information away from the cell body?

Epidermis DOES NOT contain Blood vessels

(M) = movement   Epi = secretion

Thought Questions

16. List the four major types of human body tissues and the distinguishing characteristics of each. Of what benefit does the human body contain all four types rather than a single type?

Epithelial - Skin   muscular - skeletal
Nervous - Brain/spine
connective - Blood

17. List five types of epithelial tissue and the distinguishing characteristics of each.

simple, stratified, pseudostratified, squamous, cuboidal

18. Your lab instructor gives you a slide containing prepared muscle tissue. How would you identify the type of muscle tissue located on the slide? By the layers, shape, size

19. Why might an injury to a bone have a faster recovery/healing time when compared to a muscle or nerve cell?

constantly reproduces; has more stress put on bone. when kept in constant motion, bone heals faster/stronger

McGraw-Hill Access Science Website

An online encyclopedia of science and technology that provides information, including videos, that can enhance the laboratory experience.

Find even more resources to learn the chapter concepts including animations, tutorial videos, and interactive practice questions.

www.accessscience.com

7

Organization of the Body

Learning Outcomes

7.1 Human Torso Model
- State and give a function for organs in the mouth, thoracic cavity, neck region, and abdominal cavity.

Prelab Question: State the path of food from the mouth to the anus.

7.2 External Anatomy of the Fetal Pig
- State four directive terms used when dissecting.
- Describe the external anatomy of a fetal pig.
- Tell how you would identify the gender of a fetal pig.

Prelab Question: Aside from the mouth, name two orifices of the human body.

7.3 Oral Cavity and Pharynx
- Describe the general anatomy of the oral cavity and pharynx.

Prelab Question: Explain why food does not enter the trachea.

7.4 Thoracic and Abdominal Incisions
- For those doing fetal pig dissections, describe the proper procedure for making incisions.

Prelab Question: Why is it important to point scissors up and not down?

7.5 Neck Region
- Describe and state a function for the thymus gland, the larynx, the trachea, and the thyroid gland.

Prelab Question: When might a physician be sure to check the size of the thyroid gland if a patient complains of feeling tired?

7.6 Thoracic Cavity
- Describe the position of the diaphragm.
- Describe the anatomy of the thoracic cavity, paying particular attention to the path of air.

Prelab Question: Which is more posterior, the lungs or the liver? Explain.

7.7 Abdominal Cavity
- Describe the organs of the abdominal cavity, paying particular attention to the organs of the digestive system, the liver, and the gallbladder.
- Describe the organs of the urinary system and also the male and female reproductive systems.

Prelab Question: What attribute accounts for the small intestine being called the small intestine?

Application for Daily Living: **The Systems Work Together**

Introduction

This laboratory will give you an opportunity to observe the organs of the fetal pig, a mammal, just as you are a mammal. You will be interested to know that investigators are in the process of genetically modifying pigs so that fetal pig organs can be used as transplant organs in humans.

Your instructor will decide whether you are to follow the directions for dissecting the fetal pig yourself or whether you are to observe the organs in pigs that have already been dissected. Also, virtual dissection of a human cadaver is possible using *Anatomy and Physiology Revealed*, available from McGraw-Hill Education at www.aprevealed.com.

We begin this laboratory by observing the organs to be dissected or observed in a pig by identifying the same organs in a human torso model.

7.1 Human Torso Model

As you know, most organs of the human body function as part of a particular system. Examination of the human torso model (Fig. 7.1) will allow you to learn the path of food in the digestive system, the path of air in the respiratory system, and the path of urine in the urinary system.

Figure 7.1   Human internal organs.

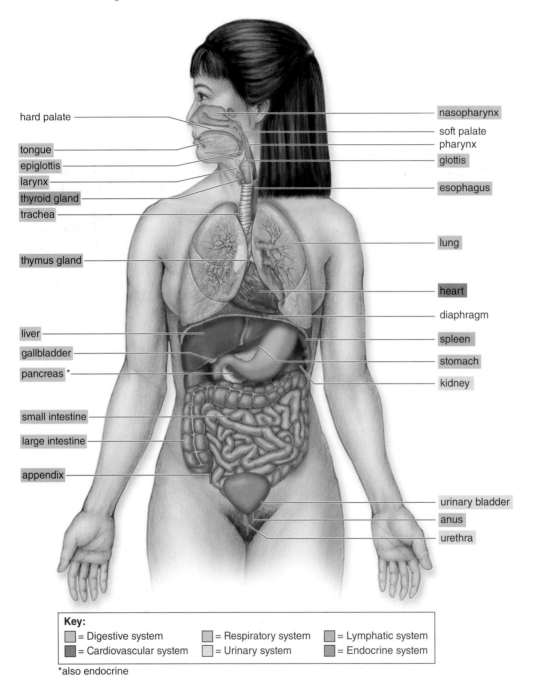

hard palate

tongue
epiglottis
larynx
thyroid gland
trachea

thymus gland

liver
gallbladder
pancreas *

small intestine

large intestine

appendix

nasopharynx
soft palate
pharynx
glottis

esophagus

lung

heart
diaphragm

spleen

stomach

kidney

urinary bladder
anus
urethra

Key:
☐ = Digestive system   ☐ = Respiratory system   ☐ = Lymphatic system
■ = Cardiovascular system ☐ = Urinary system   ☐ = Endocrine system

*also endocrine

Digestive System

To trace the path of food in the digestive system, identify these organs in the human torso model:

- **mouth:** teeth chew the food and the tongue forms a bolus (ball of food). Note the roof of the mouth consisting of the hard palate and soft palate.

- **esophagus:** conducts food to the stomach
- **stomach:** stores food, some digestion
- **small intestine:** digests food to small molecules that are absorbed into the body
- **large intestine:** absorbs water; stores undigested remains until feces are expelled from the anus

Accessory Organs of Digestion

These organs assist the digestive organs but never contain food:

- **liver:** produces bile, an emulsifier of fat
- **gallbladder:** stores bile
- **pancreas:** produces digestive enzymes that function in the small intestine

Respiratory System

To trace the path of air in the respiratory system, identify the

- **nasal passages:** in nose
- **nasopharynx:** takes air to pharynx
- **pharynx:** back of throat
- **glottis:** opening to trachea, closed by **epiglottis**
- **larynx:** (voice box) top part of trachea
- **trachea:** conducts air to bronchi
- **bronchi:** conducts air to lungs

Urinary System

To trace the path of urine, identify the

- **kidneys:** produce urine
- **urinary bladder:** stores urine
- **urethra:** conducts urine out of the body

Other Organs

Other organs include the

- **heart,** which is part of the cardiovascular system, which also contains blood vessels.
- **thyroid gland,** which produces the hormone thyroxin, and **pancreas;** which produces the hormones insulin and glucagon; the thyroid gland and pancreas which are part of the endocrine system.
- **thymus gland, spleen,** and **appendix,** which are part of the lymphatic system because they contain lymphatic tissue. The spleen also purifies the blood.

7.2 External Anatomy of the Fetal Pig

Perspective is important in dissections. If an animal is lying facedown, *top* and *bottom* mean something different than when the animal is lying on its back. For this reason, this manual uses the following terms, which are identified in Figure 7.2*a*:

Dorsal: a body part located toward the back
Ventral: a body part located toward the front
Anterior: a body part located toward the head
Posterior: a body part located toward the rear

Preparation of Pig for Dissection

1. Place the fetal pig on its back in the dissecting pan.
2. Tie a cord around one forelimb, and then bring the cord around underneath the pan to fasten back the other forelimb.
3. Spread the hindlimbs in the same way.

Figure 7.2 External anatomy of the fetal pig.

a. Directional terms and body regions. **b, c.** The sexes can be distinguished by the external genitals.

a. Body region

b. Female

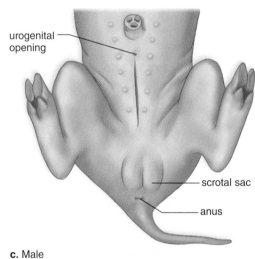

c. Male

Observation: External Anatomy

These observations can be done by all students.

Body Regions

1. Externally, observe the following body regions: the rather large head; the short, thick neck; the cylindrical trunk with two pairs of appendages (forelimbs and hindlimbs); and the short tail (Fig. 7.2*a*).

> ⚠ **Latex gloves** Wear protective latex gloves when handling preserved animal organs. Use protective eyewear and exercise caution when using sharp instruments during this laboratory. Wash hands thoroughly upon completion of this laboratory.

Umbilical Cord

1. Fetal pigs have an umbilical cord arising from the ventral portion of the abdomen. Fetal pigs have not been born yet, and this is why they have a prominent umbilical cord. The function of the umbilical cord is to

bring oxygen and nutrients from the mother's circulation to the fetus and return oxygen and waste to the mother's circulation.

Nipples and Hair

Like all **mammals,** pigs and humans have mammary glands and hair.

1. Nipples, the external openings of the **mammary glands,** are not an indication of sex, because both males and females possess them.
2. Pigs have a few whiskers, the only indication of hair.

Anus and External Genitals

1. The anus, an opening under the tail, is a part of what system in the body? _____
2. Females have a **urogenital opening,** just anterior to the anus, and a small, fleshy **urogenital papilla** projecting from the urogenital opening (Fig. 7.2*b*).
3. Males have a urogenital opening just posterior to the umbilical cord. The duct leading to it runs forward from between the legs in a long, thick tube, the **penis,** which can be felt under the skin. In males, the urinary system and the genital system are always joined (Fig. 7.2*c*).

7.3 Oral Cavity and Pharynx

All students will use Figure 7.3 to make these observations:

The **oral cavity** is the space in the **mouth** that contains the **tongue** and the **teeth.** What system of the body

includes the tongue and teeth? _____

The **hard palate** is the ridged roof of the mouth that separates the oral cavity from the nasal passages. The **soft palate** is a smooth region posterior to the hard palate. An extension of the soft palate—the uvula—hangs down into the throat in humans. (A pig does not have an uvula.)

Observation: Oral Cavity and Pharynx

If you are doing a pig dissection, follow these directions. If you are observing a pig that has already been dissected or are doing a virtual dissection, answer the questions and proceed to Section 7.5.

Oral Cavity

1. Insert a sturdy pair of scissors into one corner of the specimen's mouth, and cut posteriorly (toward the hind end) for approximately 4 cm. Repeat on the opposite side.
2. Place your thumb on the tongue at the front of the mouth, and gently push downward on the lower jaw. This will tear some of the tissue in the angles of the jaws so that the mouth will remain partly open (Fig. 7.3).
3. Note small, underdeveloped teeth in both the upper and lower jaws. Other embryonic, nonerupted teeth may also be found within the gums. The teeth are used to chew food.

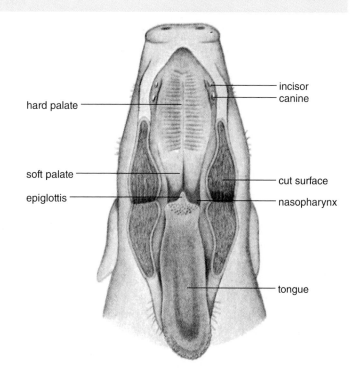

Figure 7.3 Oral cavity of the fetal pig.
The roof of the oral cavity contains the hard and soft palates. The tongue lies above the floor of the oral cavity.

4. Examine the tongue, partly attached to the lower jaw region but extending posteriorly and attached to a bony structure at the back of the oral cavity (Fig. 7.3).
5. Locate the hard and soft palates (Fig. 7.3).

Pharynx

1. Push down on the tongue until you open the jaws far enough to see a slightly pointed flap of tissue pointing dorsally (toward the back) (Fig. 7.3). This flap is the **epiglottis,** which covers the glottis. The glottis leads to the trachea (Fig. 7.4).
2. Posterior and dorsal to the glottis, find the opening into the esophagus. Note the proximity of the glottis and the opening to the esophagus. Each time the pig—or a human—swallows, the epiglottis shifts position, closing the glottis to keep food and fluids from going into the lungs via the trachea.
3. Insert a blunt probe into the glottis, and it enters the trachea. Remove the probe, insert it into the esophagus, and note the position of the esophagus beneath, or dorsal to, the trachea.
4. Make a midline cut in the soft palate from the epiglottis to the hard palate. Then make two lateral cuts at the edge of the hard palate.
5. Posterior to the soft palate, locate the openings to the nasal passages.
6. Explain why it is correct to say that the air and food passages cross in the pharynx. _____

_____

Figure 7.4 Air and food passages in the fetal pig.
The air and food passages cross in the pharynx.

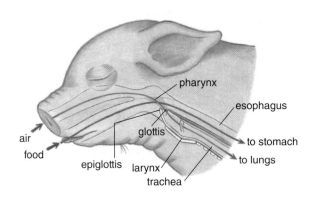

7.4 Thoracic and Abdominal Incisions

Notice that there is a slash in the right neck region of preserved pigs, indicating the site of blood drainage. A red latex solution was injected into the **arterial system,** and a blue latex solution was injected into the **venous system** of the pig. Therefore, when a vessel appears red, it is an artery; and when a vessel appears blue, it is a vein. Do not confuse this color pattern with circulatory diagrams that differentiate O_2-rich blood flow and O_2-poor blood flow by using red- and blue-colored vessels, respectively.

If you are observing a pig that has been dissected or doing a virtual dissection, proceed to Section 7.5.

If you are doing a pig dissection, consult Figure 7.5 and make the following incisions with scissors pointing up and never down to expose the thoracic and abdominal cavities.

Incisions

1. Cut anteriorly up from the **diaphragm,** a muscle that separates the thoracic cavity from the abdominal cavity, until you reach the hairs in the throat region.
2. Make two lateral cuts, one on each side of the midline incision anterior to the forelimbs, taking extra care not to damage the blood vessels around the heart.
3. Make two lateral cuts, one on each side of the midline just posterior to the forelimbs and anterior to the diaphragm, following the ends of the ribs. Pull back the flaps created by these cuts to expose the **thoracic cavity.** List the organs you find in the thoracic cavity.

Figure 7.5 Ventral view of the fetal pig indicating incisions.

These incisions are to be made preparatory to dissecting the internal organs. The incisions are numbered here in the order they should be done.

nostril

eye

external ear

Right forelimb

wrist

1

2

digit

Left forelimb

3

elbow

6

4

diaphragm

umbilical cord

7

knee

5

Right hindlimb

Left hindlimb

anus

ankle

tail

4. With scissors pointing up, cut posteriorly from the diaphragm to the umbilical cord.
5. Make a flap containing the umbilical cord by cutting a semicircle around the cord and by cutting posteriorly to the left and right of the cord.
6. Make two cuts, one on each side of the midline incision posterior to the diaphragm. Examine the diaphragm, attached to the chest wall by radially arranged muscles. The central region of the diaphragm, called the **central tendon,** is a membranous area.
7. Make two more cuts, one on each side of the flap containing the umbilical cord and just anterior to the hindlimbs. Pull back the side flaps created by these cuts to expose the **abdominal cavity.**
8. Lifting the flap with the umbilical cord requires cutting the **umbilical vein.** Before cutting the umbilical vein, tie a thread on each side of where you will cut to mark the vein for future reference.

7.5 Neck Region

All students will use Figure 7.6 to make these observations:

1. The **larynx,** or voice box, is part of the **trachea,** or windpipe, which leads to the lungs. The glottis is the opening to the trachea in the pharynx. The trachea is held open by cartilaginous rings. Why is that

   advantageous? _____

2. Unlike humans, the thymus gland is located in the neck region of a fetal pig. The thymus gland is a part of the lymphatic system. Certain white blood cells called T (for thymus) lymphocytes mature in the thymus gland and help fight disease.

3. The thyroid gland is ventral to the trachea. It secretes hormones that travel in the blood and act upon other body cells, causing their metabolism to speed up. These hormones (e.g., thyroxine) regulate the rate at which metabolism occurs in cells.

4. The **esophagus** opens into the pharynx but it travels through the neck region and through the diaphragm to reach the stomach. Is the esophagus located ventral to or dorsal to the trachea?

Figure 7.6   The neck region of the fetal pig.
The trachea, including the larynx, is in the neck region. Various other organs are also in the neck region.

Observation: Neck Region

If you are doing a pig dissection or are observing a pig that has been dissected, follow these directions. If you are not doing either, proceed to Section 7.6.

Use Figure 7.6 as a guide, but keep all the flaps on your pig so you can close the thoracic and abdominal cavities at the end of the laboratory session. **Never remove any organs from a fetal pig;** you will want to review them later.

Thymus Gland

1. Move the skin apart in the neck region just below the hairs mentioned earlier. If necessary, laterally cut the body wall to make flaps. You will most likely be viewing exposed muscles.
2. Cut through and clear away muscle to expose the thymus gland, a diffuse gland that lies among the muscles. The thymus is particularly large in fetal pigs, because their immune systems are still developing.

Trachea and Esophagus

1. Probe down into the deeper layers of the neck. Medially (toward the center), beneath several strips of muscle, you will find the hard-walled larynx, the upper part of the trachea. Dorsal to the trachea, find the esophagus.
2. Open the mouth and insert a probe into the glottis and esophagus from the pharynx to better understand the orientation of these two organs.

Thyroid Gland

Locate the thyroid gland just posterior to the larynx, lying ventral to the trachea.

7.6 Thoracic Cavity

All students will use Figure 7.7 to make these observations:

The body cavity of mammals, including humans, is divided by the **diaphragm** (horizontal sheet of muscle and connective tissue) into the **thoracic cavity** (above) and the **abdominal cavity** (below). These organs are easily observable in the thoracic cavity:

1. The **heart** pumps the blood into attached arteries, blood vessels that take blood away from the heart. The heart is surrounded by the **pericardial membrane,** which prevents friction between organs. The heart is a

   part of which system in the body? _____

2. The trachea divides into the **bronchi,** which enter the **lungs.** The lungs are enclosed by **pleural**

   membrane. The lungs are a part of which system in the body? _____

3. The lungs carry on gas exchange with the external environment and with the blood. The right lung has four lobes and the left lung has three lobes.

Observation: Thoracic Cavity

If you are doing a pig dissection or are observing a pig that has been dissected, follow these directions. If you are doing a virtual dissection, proceed to Section 7.7.

Heart and Lungs

1. If you have not yet done so, fold back the chest wall flaps. To do this, you will need to tear the thin membranes that divide the thoracic cavity into three compartments: the left **pleural cavity** containing the left lung, the right pleural cavity containing the right lung, and the **pericardial cavity** containing the heart.
2. Examine the lungs. Locate the four lobes of the right lung and the three lobes of the left lung. The trachea, dorsal to the heart, divides into the bronchi, which enter the lungs. Later, when the heart is removed, you will be able to see the trachea and bronchi.

Summary: Respiratory System

Trace the path of air in the respiratory system from pharynx to lungs. _____

_____

Figure 7.7 Internal anatomy of the fetal pig.

Most of the major organs are shown in this photograph. For the sake of visibility, the incisions have not preserved the entire diaphragm. The stomach has been removed. The spleen, gallbladder, and pancreas are not visible. **Do not remove any organs or flaps from your pig.**

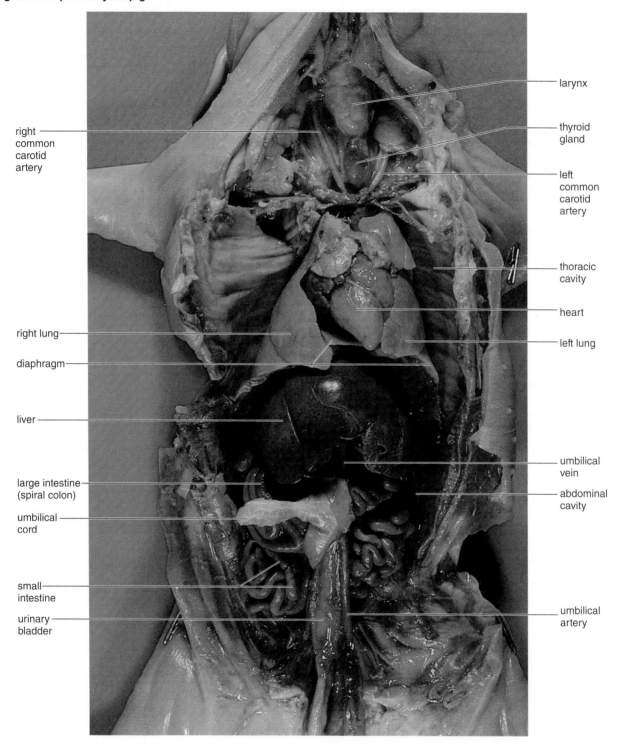

larynx

thyroid gland

right common carotid artery

left common carotid artery

thoracic cavity

heart

right lung

left lung

diaphragm

liver

umbilical vein

large intestine (spiral colon)

abdominal cavity

umbilical cord

small intestine

umbilical artery

urinary bladder

7.7 Abdominal Cavity

All students will use Figure 7.7 to make these observations:

The abdominal wall and organs are lined by a connective tissue membrane called **peritoneum.** Double-layered sheets of peritoneum, called **mesenteries,** project from the body wall and support the organs.

1. The abdominal cavity contains organs of the **digestive system,** including the stomach, small intestine, and large intestine. The **stomach** stores food and has numerous gastric glands that secrete gastric juice, which digests protein. The **small intestine** is the part of the digestive tract that receives secretions from the pancreas and gallbladder. Ducts from these organs enter the **duodenum,** the first part of the small intestine. Besides being an organ for the digestion of all components of food—carbohydrate, protein, and fat—the small intestine absorbs the products of digestion: glucose, amino acids, glycerol, and fatty acids. The **large intestine** is the part of the digestive tract that absorbs water and prepares feces for defecation at the anus. The main portion of the large intestine is the **colon,** which first runs anteriorly and then horizontally before turning posteriorly again. In the pelvic region, the **rectum** is the last portion of the large intestine. The rectum leads to the **anus.**

2. The abdominal cavity also contains the accessory organs of digestion. These organs send substances to the digestive tract and never contain food. The **liver,** the largest organ in the abdomen (Fig. 7.7), performs numerous vital functions, including disposing of worn-out red blood cells, producing bile, storing glycogen, maintaining the blood glucose level, and producing blood proteins. The **gallbladder** stores and releases bile, which aids the digestion of fat. A duct from the gallbladder takes bile to the small intestine. Bile is an emulsifier that breaks up fat into smaller fat droplets, enabling better accesss by digestive enzymes. The **pancreas** is both an exocrine and an endocrine gland. As an exocrine gland, it produces and secretes pancreatic juice, which digests all the components of food in the small intestine. Both bile and pancreatic juice enter the duodenum by way of ducts. As an endocrine gland, the pancreas secretes the hormones insulin and glucagon into the bloodstream.

Observation: Abdominal Cavity

If you are doing a pig dissection or observing a pig that has already been dissected, follow these directions. If you are doing a virtual dissection, answer the questions.

Liver

1. If your particular pig is partially filled with dark, brownish material, take your animal to the sink and rinse it out. This material is clotted blood. Consult your instructor before removing any red or blue latex masses, because they may enclose organs you will need to study.
2. Locate the liver, a large, brown organ. Its anterior surface is smoothly convex and fits snugly into the concavity of the diaphragm.
3. Name several functions of the liver. _____

_____

Stomach

1. Push aside and identify the stomach, a large sac dorsal to the liver on the left side.
2. Locate the point near the midline of the body where the esophagus penetrates the diaphragm and joins the stomach.

Small Intestine

1. Look posteriorly where the stomach makes a curve to the right and narrows to join the anterior end of the small intestine. The first part of the small intestine is called the duodenum.
2. From the duodenum, the small intestine runs posteriorly for a short distance and is then thrown into an irregular mass of bends and coils held together by a common mesentery.

3. The small intestine is a part of what system? _____

What is its function? _____

Gallbladder and Pancreas

1. Locate the **bile duct,** which runs in the mesentery stretching between the liver and the duodenum. Find the gallbladder, embedded in the liver on the underside of the right lobe. It is a small, greenish sac.

2. Lift the stomach and locate the pancreas, the light-colored, diffuse gland lying in the mesentery between the stomach and the small intestine. The pancreas has a duct that empties into the duodenum of the small intestine.

3. What is the function of the gallbladder? _____

4. How does the pancreas assist the digestion of food? _____

Large Intestine

1. Locate the distal (far) end of the small intestine, which joins the large intestine posteriorly, in the left side of the abdominal cavity (right side in humans). At this junction, note the cecum, a blind pouch.

2. Follow the main portion of the large intestine, known as the colon, as it runs from the point of juncture with the small intestine into a tight coil (spiral colon), out of the coil anteriorly, and then posteriorly again along the midline of the dorsal wall of the abdominal cavity. In the pelvic region, the rectum is the last portion of the large intestine. The rectum leads to the anus.

3. The large intestine is a part of what system? _____

4. What is the function of the large intestine? _____

Summary: Digestive System

Trace the path of food from the mouth to the anus. _____

_____

Kidneys and Associated Structures

1. The large, paired kidneys (Fig. 7.8) are reddish organs covered by **peritoneum,** a membrane that lines the abdominal cavity. Clean the peritoneum away from one of the kidneys, and study it more closely.

2. Locate the ureters, which leave the kidneys and run posteriorly under the peritoneum.

3. Clean the peritoneum away, and follow a ureter to the **urinary bladder,** which normally lies in the posterior ventral portion of the abdominal cavity. The urinary bladder is on the inner surface of the flap of tissue to which the umbilical cord was attached.

4. The **urethra,** which arises from the bladder posteriorly, runs parallel to the rectum. Follow the urethra until it passes from view into the ring formed by the pelvic girdle.

Summary: Urinary System

Trace the path of urine from the kidneys to the urethral opening. _____

_____

Ovaries and Associated Structures in Female Pigs

1. Locate the paired ovaries, small bodies suspended from the peritoneal wall in mesenteries, posterior to the kidneys (Fig. 7.8).

2. Closely examine one ovary. Note the small, short, coiled **oviduct,** sometimes called the fallopian tube. The oviduct does not attach directly to the ovary but ends in a funnel-shaped structure with fingerlike processes (fimbriae) that partially enclose the ovary.

Figure 7.8   Female urinary and reproductive systems of the fetal pig.
In the adult female, the urinary system and the reproductive system are separate. In the fetus, the vagina joins the urethra just before the urogenital sinus.

kidney

ureter

ovarian vein

ovary

uterine horn

body of uterus

vagina

urinary bladder

urethra

urogenital sinus

umbilical arteries

urogenital papilla

umbilical cord

3. Locate the **uterine horns.** (Do not confuse the uterine horns with the oviducts; the latter are much smaller and are found very close to the ovaries.)
4. Find the median body of the uterus located at the joined posterior ends of the uterine horns.
5. Separate the hind limbs of your specimen, and cut down along the midventral line. The cut will pass through muscle and the cartilaginous pelvic girdle. With your fingers, spread the cut edges apart, and use blunt dissecting instruments to separate connective tissue.
6. Note three ducts passing from the body cavity to the animal's posterior surface. One of these is the urethra, which leaves the bladder and passes into the **urogenital sinus.** The urethra is a part of the urinary system. The most dorsal of the three ducts is the **rectum,** which passes to its own opening, the **anus.** The rectum and anus are part of the digestive system, not the reproductive system.
7. Find the vagina, located dorsally to the urethra. The vagina is the birth canal and is also the organ of copulation. Anteriorly, it connects to the uterus, and posteriorly it enters the urogenital sinus. This sinus is absent in adult humans and several other female mammals.

Testes and Associated Structures in Male Pigs

1. Locate the opening of the left inguinal canal, which leads to the left scrotal sac (Fig. 7.9).
2. Expose the canal and sac by making an incision through the skin and muscle layers from a point over this opening back to the left scrotal sac.
3. Open the sac, and find the testis. Note the much-coiled tubule—the epididymis—that lies alongside the testis. This is continuous with the vas deferens, which passes back toward the abdominal cavity.
4. Trace a vas deferens as it loops over an umbilical artery and ureter and unites with the urethra dorsally at the posterior end of the urinary bladder. The urethra passes into the penis and ends at the urogenital opening (see Fig. 7.2c).

Figure 7.9   Male urinary and reproductive systems of the fetal pig.
In males, the urinary system and the reproductive system are joined. The vasa deferentia (sing. vas deferens) enter the urethra, which also carries urine.

If so directed by your instructor:

1. Cut through the ventral skin surface just posterior to the umbilical cord. This will expose the rather undeveloped penis, which extends from this point posteriorly toward the anus. The central duct of the penis is the urethra.

2. Lay the penis to one side, and then cut down through the ventral midline, laying the legs wide apart in the process (Fig. 7.9). The cut will pass between muscles and through pelvic cartilage (bone has not developed yet). Do not cut any of the ducts or tracts in the region.

3. You will now see the urethra passing ventrally above the rectum. It is associated with certain accessory glands:

   a. Bulbourethral glands, about 1 cm in diameter, lie laterally and well back toward the anal opening.

   b. The prostate gland, about 4 mm across and 3 mm thick, is located on the dorsal surface of the urethra, just posterior to the juncture of the bladder with the urethra. It is often difficult to locate and is not shown in Figure 7.9.

   c. Small, paired seminal vesicles may be seen on either side of the prostate gland.

4. Trace the urethra as it leaves the bladder. It proceeds posteriorly, but when it nears the posterior end of the body, it turns rather abruptly anterioventrally and runs forward just under the skin of the midventral body wall, where you have just dissected it. This latter portion of the urethra is, then, within the penis.

5. Now you should also be able to see the entrance of the vasa deferentia into the urethra. If necessary, dissect these structures free from surrounding tissue, and expose the point of entrance of these ducts into the urethra near the location of the prostate gland. In males, the urethra transports sperm, as well as urinary wastes from the bladder.

The Systems Work Together

This laboratory has involved a number of different systems, including the respiratory system, the cardiovascular system, and the digestive system. Can one of these systems malfunction without negatively affecting the other systems? Likely not. For example, if you climb a high mountain and can't breathe adequately, the cardiovascular system will not be able to deliver oxygen to your cells (Fig. 7.10). Or, if you go on a hunger strike, the cardiovascular system will not be able to deliver glucose to your cells. Without oxygen and glucose, even your heart will not be able to continue beating. Blood vessels not only deliver oxygen and glucose to the heart; they also take these molecules to the lungs and the walls of the digestive tract.

In the body as in the environment, everything is connected to everything else. The systems work together under the direction of the nervous and endocrine systems as a harmonious whole so that your cells and you can continue to function as you should.

Figure 7.10   The systems work together.
The cardiovascular system receives nutrients from the digestive tract and oxygen from the respiratory system, which it delivers to the cells. It also delivers CO_2 and other wastes from the cells to the respiratory system and urinary system respectively.

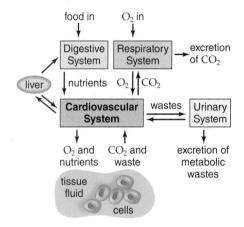

Storage of Pigs

1. Before leaving the laboratory, place your pig in the plastic bag provided.
2. Expel excess air from the bag, and tie it shut.
3. Write your name and section on the tag provided, and attach it to the bag. Your instructor will indicate where the bags are to be stored until the next laboratory period.
4. Clean the dissecting tray and tools, and return them to their proper location.
5. Wipe off your goggles. Wash your hands.

_____ 1. In the fetal pig, what sex has a urogenital opening beneath the papilla just superior to the anus?

_____ 2. What two anatomical features do all mammals, including pigs and humans, have?

_____ 3. The esophagus connects the pharynx with which organ?

_____ 4. What is the hard portion of the roof of the mouth called?

_____ 5. What is the opening to the trachea called?

_____ 6. Name the largest organ in the abdominal cavity.

_____ 7. What structure separates the thoracic cavity from the abdominal cavity?

_____ 8. Name the structure just dorsal to the thyroid gland.

_____ 9. What structure covers the glottis?

_____ 10. If a probe is placed through the glottis, it will enter what structure?

_____ 11. The heart is located in what major cavity?

_____ 12. What other major organs, surrounded by the pleural membranes, are in the same cavity?

_____ 13. The stomach connects to what part of the small intestine?

_____ 14. The pancreas belongs to what two systems of the body?

_____ 15. Where do air and food passages cross one another?

_____ 16. What organ releases bile?

Thought Questions

17. What difficulty would probably arise if a person were born without an epiglottis?

18. The small intestine exists as a series of folds and coils. What might be the advantage of such a configuration?

19. Difficulties maintaining blood glucose level, bile production, and the production of blood proteins might be associated with problems in what organ?

20. How do the male and female reproductive systems differ?

BIOLOGY

Find even more resources to learn the chapter concepts including animations, tutorial videos, and interactive practice questions.

McGraw-Hill Access Science Website

An online encyclopedia of science and technology that provides information, including videos, that can enhance the laboratory experience.

www.accessscience.com

8

Cardiovascular System

Learning Outcomes

8.1 The Heart
- Name the four chambers of the heart and the blood vessels attached to these chambers.
- Trace the path of blood through the heart, naming the chambers and the valves in proper order.
- Describe the conduction system of the heart, including the nodes that control the heartbeat.

Prelab Question: Which side of the heart pumps O_2-poor blood and which side pumps O_2-rich blood?

8.2 Heartbeat and Blood Pressure
- Describe the heartbeat, including the sounds that occur when the heart beats.
- Relate blood pressure to the beat of the heart.
- Show blood pressure rises with exercise and offer an explanation.

Prelab Question: Relate normal blood pressure (120/80) to the beat of the heart.

8.3 The Blood and Blood Flow
- Distinguish between red blood cells and the various types of white blood cells.
- Explain the ABO system of blood typing and how blood type is determined.
- Describe the structure and function of arteries, veins, and capillaries.
- Name the major blood vessels and trace the path of blood in both the pulmonary and systemic circuits.

Prelab Question: Relate the axiom "structure suits function" to the function of the different types of blood cells.

Application for Daily Living: **High Blood Pressure**

Introduction

The heart and blood vessels form the cardiovascular system. The heart is a double pump that (1) keeps the blood flowing in one direction—blood flows away from and then back to the heart; (2) keeps O_2-poor blood separate from O_2-rich blood; and (3) creates blood pressure, which moves the blood through the blood vessels. The blood vessels, on the other hand, (1) transport blood and its contents, (2) serve the needs of the body's cells by carrying out exchanges with them, and (3) direct blood flow to those systemic tissues that most require it at the moment.

We will make note of these functions again as we first study the structure and beat of the heart and how its beat brings about blood pressure. Following that, we concentrate on the composition of blood, how it is typed, and how it circulates in the body.

A thoracic cavity contains the heart and lungs.

8.1 The Heart

The heart has right and left sides divided by the **septum.** To tell the right from the left side, position Figure 8.1 so it corresponds to your body. There are four **chambers:** two upper, thin-walled atria and two lower, thick-walled ventricles. The heart has valves that keep the blood flowing in one direction; special muscles and tendons secure the valves to prevent backflow from the ventricles to the atria. The right side of the heart sends blood to the lungs, and the left side sends blood into the body. Therefore, the heart is called a double pump.

Observation: External Anatomy of the Heart

1. In a heart model and/or a sheep heart, identify the **right atrium** and its attached blood vessels, the superior and inferior **venae cavae** (Fig. 8.1). The superior vena cava and the inferior vena cava return blood from the head and body, respectively, to the right atrium. Is the blood that enters the right atrium, O_2-poor or O_2-rich? <u>O2 Poor</u> Explain. _____

    _____

2. Identify the **right ventricle** and its attached blood vessel, the **pulmonary trunk.** The pulmonary trunk leaves the ventral side of the heart from the top of the right ventricle and then passes diagonally forward, before branching into the **right** and **left pulmonary arteries.**

3. Identify the **left atrium** and its attached blood vessels, the left and right **pulmonary veins.** The pulmonary veins return blood from the lungs to the left atrium. Is the blood that enters the left atrium O_2-poor or O_2-rich? <u>O2 Rich</u> Explain. <u>Coming straight to Lungs</u>

    _____

4. Identify the **left ventricle** and its attached blood vessel, the **aorta,** which arises from the anterior end of the left ventricle, just dorsal to the origin of the pulmonary trunk. The aorta soon bends to the animal's left as the aortic arch. The aorta carries blood to the body proper.

Figure 8.1 External heart anatomy.
a. Human heart. **b.** Sheep heart.

a. Human heart

b. Sheep heart

5. Identify the **coronary arteries** and the **cardiac veins,** which service the needs of the heart wall. The coronary arteries branch off the aorta as soon as it leaves the heart and appear on the surface of the heart. The cardiac veins, also on the surface of the heart, join and then enter the right atrium through the coronary sinus on the dorsal side of the heart.

Observation: Internal Anatomy of the Heart

Remove the ventral half of the human heart model (Fig. 8.2) and/or sheep heart. You are going to remove the top portion of the heart to achieve a view similar to Figure 8.3*b*. Position the heart same as it is in Figure 8.2. Using a sharp scalpel, cut through the left atrium and the left ventricle to the apex of the heart, being sure your cut is deep enough to reach the septum. Now position the heart so that the apex is uppermost. Using a sharp scalpel, cut from the apex of the heart through the right atrium, being sure your cut is deep enough to reach the septum of the heart.

> ⚠️ **Latex gloves** Wear protective latex gloves when handling preserved animal organs. Use protective eyewear and exercise caution when using sharp instruments during this laboratory. Wash hands thoroughly upon completion of this laboratory.

1. Identify the four chambers of the heart in longitudinal section: right atrium, right ventricle, left atrium, and left ventricle. *Label Figure 8.2.*

2. Which ventricle is more muscular? _left, Has to push blood all over body_

Why is this appropriate? _____

3. Find the **right atrioventricular** (tricuspid) valve, located between the right atrium and the right ventricle.

4. Find the **left atrioventricular** (bicuspid or mitral) valve, located between the left atrium and the left ventricle.

5. Find the **pulmonary semilunar** valve, located in the base of the pulmonary trunk.

6. Find the **aortic semilunar** valve, located in the base of the aorta. What is the function of the

heart valves? _____

7. Note the **chordae tendineae** ("heartstrings") that hold the atrioventricular valves in place while the heart contracts. These extend from the papillary muscles. The chordae tendineae prevent the atrioventricular valves from inverting into the atria when the ventricles contract.

8. If inspecting a sheep heart, insert a probe into the various vessels from the outside into the heart chambers. Identify the vessel and the chamber.

Figure 8.2   Internal heart anatomy.
a. Human heart. **b.** Sheep heart.

a. Human heart

b. Sheep heart

Laboratory 8   Cardiovascular System

Summary

To demonstrate that O_2-poor blood is kept separate from O_2-rich blood, trace the path of blood from the right side of the heart to the aorta by filling in the following blanks.

Venae Cavae

R. Atrium

tricuspid valve

Right ventricle

pulm semilunar valve

pulm. teunle

Lungs

Lungs

pulm. veins

L Atrium

Bicuspid valve

L ventricle

aortic valve
semilunar

Aorta

Which side of the heart (right or left) pumps O_2-poor blood? R

Which side pumps O_2-rich blood?

L

Physiology of the Heart

The heartbeat is controlled by nodal tissue. The SA node is the pacemaker of the heart because it initiates the heartbeat and sends an excitation impulse every 0.85 seconds, causing the atria to contract. After the impulse reaches the **AV node,** it passes into large fibers and, thereafter, spreads out by way of the smaller **Purkinje fibers** (Fig. 8.3a). These fibers signal the ventricles to contract.

With the contraction of any muscle, including the heart, electrolyte changes occur that can be detected by electrical recording devices. Therefore, it is possible to study the heartbeat by recording voltage changes that occur when the heart contracts. The record that results is called an electrocardiogram (ECG). The first wave in the electrocardiogram, called the P wave, occurs prior to the excitation and contraction of the atria. The second wave, the QRS wave, occurs prior to ventricular excitation and contraction. The third wave, the T wave, is caused by the recovery of the ventricles. (Atrial relaxation is not apparent in an ECG.) Examination of an ECG indicates whether the heartbeat has a normal or an irregular pattern (Fig. 8.3b).

1. Note the SA (sinoatrial) node in Figure 8.3a.
2. Explain Figure 8.3b by answering these questions: Why is an arrow drawn between the SA node and the P wave?

_____

Why is an arrow drawn between the AV node and the QRS wave?

_____

The voltage changes in an ECG are related to the

_____ .

Figure 8.3 Control of the heartbeat.
a. The SA node sends out a stimulus that causes the atria to contract. When this stimulus reaches the AV node, it signals the ventricles to contract by way of the atrioventricular bundle and Purkinje fibers. **b.** A normal ECG indicates that the heart is functioning properly. The P wave indicates that the atria are about to contract; the QRS wave indicates that the ventricles are about to contract; and the T wave indicates that the ventricles are recovering from contraction.

1. Electrodes will be placed on the wrists and ankles, so the subject should remove any jewelry from these areas. The subject then lies down on a cot or table close to the electrocardiograph.
2. Clean the electrode placement areas with alcohol swabs and apply a small quantity of electron cream to these areas.
3. Similarly spread electrode cream on the inner surfaces of four electrode plates before attaching one to each of the four sites on the subject. The electrode plate on the right ankle is the grounding system.
4. A selector switch allows various combinations of electrodes to be activated. Attach the leads from the electrodes to the corresponding cables of the lead selector switch.

    Lead I measures the potential difference between the right wrist and the left wrist.

    Lead II measures the potential difference between the right wrist and the left ankle.

    Lead III measures the potential difference between the left wrist and the left ankle.

    Attachment sites

5. Record the ECG for Lead I, Lead II, and Lead III for one minute each. Then remove the electrodes and clean the cream from the metal and skin of the subject.
6. Do the ECGs you recorded resemble that in Fig. 8.3? _____

    Explain. _____

8.2 Heartbeat and Blood Pressure

In this section, we will examine the heartbeat and blood pressure. The beat of the heart creates blood pressure.

Heartbeat

During a heartbeat, first the atria contract and then the ventricles contract. When a chamber contracts, it is called **systole;** when a chamber relaxes, it is called **diastole.** The atria and ventricles take turns being in systole:

Time	Atria	Ventricles
0.15 sec	Systole	Diastole
0.30 sec	Diastole	Systole
0.40 sec	Diastole	Diastole

Usually, there are two heart sounds with each heartbeat. The first sound (lub) is low and dull and lasts longer than the second sound. It is caused by the closure of the atrioventricular valves following atrial systole. The second sound (dub) follows the first sound after a brief pause. The sound has a snapping quality of higher pitch and shorter duration. The sound is caused by the closure of the semilunar valves following ventricle systole.

In the following procedure, you will work with a partner and use a stethoscope to listen to the heartbeat. It will not be necessary for you to count the number of beats per minute.

1. Obtain a stethoscope, and properly position the earpieces. They should point forward. Place the bell of the stethoscope on the left side of your partner's chest between the fourth and fifth ribs. This is where the apex (tip) of the heart is closest to the body wall.

2. Which of the two sounds (lub or dub) is louder? _____

3. Now switch, and your partner will determine your heartbeat.

Blood Pressure

Blood pressure is highest just after ventricular systole, and it is lowest during ventricular diastole.

Why? _____

We would expect a person to have lower blood pressure readings at rest than after exercise.

Why? _____

Experimental Procedure: Blood Pressure at Rest and After Exercise

A number of different types of digital blood pressure monitors are available, and your instructor will instruct you on how to use the type you will be using for this Experimental Procedure. The normal resting blood pressure readings for a young adult are 120/80 (systolic/diastolic), as displayed on the monitor shown in Figure 8.4.

You may work with a partner or by yourself. If working with a partner, each of you will assist the other in taking blood pressure readings. After you have noted the blood pressure readings, also note the pulse reading.

Blood Pressure at Rest

1. Reduce your activity as much as possible.
2. Use the blood pressure monitor to obtain several blood pressure readings, average them, and record your results in Table 8.1.

Blood Pressure After Exercise

1. Run in place for 1 minute.
2. Immediately use the blood pressure monitor to obtain a blood pressure reading, and record it in Table 8.1.

Figure 8.4 Measurement of blood pressure and pulse.
There are many different types of digital blood pressure/pulse monitors now available. The one shown here uses a cuff to be placed on the arm. Others use a cuff for the wrist.

| Table 8.1   Blood Pressure | |
Blood Pressure at Rest	Blood Pressure After Exercise
Partner	
Yourself	

Conclusion: Blood Pressure

- Recall that in Laboratory 1, you determined that exercise increases the heartrate. Knowing this, offer an explanation for your results. _____

- Under what conditions in everyday life would you expect the heartrate and the blood pressure to increase, even though you are not exercising? _____
  When might this be an advantage? A disadvantage? _____
  _____

8.3 The Blood and Blood Flow

Blood appears to be a somewhat viscous, homogeneous fluid. However, analysis shows it to be a fluid tissue, composed of plasma (the fluid portion) and formed elements (including the cells and platelets).

Red and White Blood Cells

There are two categories of blood cells: red blood cells (erythrocytes) and white blood cells (leukocytes). Red blood cells are smaller than white blood cells, and they lack a nucleus, which enables them to carry the maximum amount of oxygen. Red blood cells appear red because they contain the respiratory pigment hemoglobin. Each red blood cell lasts about 120 days in circulation.

White blood cells are larger than red blood cells, and they have a nucleus. The white blood cells are translucent if not stained; there are five different types (Fig. 8.5). White blood cells fight infection, and the white blood cell count is used by doctors to help diagnose diseases. The time white blood cells are in circulation is variable.

Figure 8.5 The white blood cells.
a. A neutrophil has a lobed nucleus with two to five lobes. **b.** An eosinophil has red-staining granules. **c.** A basophil has deep-blue-staining granules. **d.** A monocyte is the largest of the blood cells. **e.** A lymphocyte contains a large, round nucleus. (**a–c, e:** Magnification ×400; **d:** Magnification ×500)

a. Neutrophil

b. Eosinophil

c. Basophil   platelets

d. Monocyte

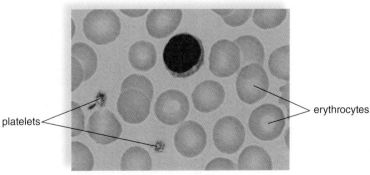

platelets   erythrocytes

e. Lymphocyte

1. Observe a prepared blood smear slide on high power, and note the biconcave (concave on both sides) red blood cells and the less numerous white blood cells. Also, differentiate the two types of cells according to their size and the presence or absence of a nucleus.
2. Complete Table 8.2.

Table 8.2 Slide of Blood	Relative Number	Relative Size	Presence of Nucleus	Function
Red blood cells				
White blood cells				

Observation: White Blood Cell Slide

1. Observe the demonstration slides under oil immersion. With the help of Figure 8.5, identify the different types of white blood cells, which have been stained with Wright stain.
2. Identify the **granular leukocytes:**
   Neutrophils have cytoplasmic granules; the nucleus is multilobed with two to five connected parts. Neutrophils are, therefore, called polymorphonuclear leukocytes.
   Eosinophils have granules that stain deep red. The nucleus is bilobed.
   Basophils have granules that stain deep blue. The nucleus is bilobed.
3. Identify the **agranular leukocytes:**
   Monocytes are the largest of the white blood cells. The nucleus varies in shape.
   Lymphocytes are usually only slightly larger than red blood cells and typically have a relatively large, round nucleus surrounded by a thin rim of cytoplasm.

ABO Blood Typing

Red blood cells have surface molecules called antigens that indicate they belong to the person. In the ABO system, the presence or absence of an A antigen and/or a B antigen determines the blood type. In this system, there are four types of blood: A, B, AB, and O. Within the plasma, there are antibodies to the antigens that are not on the red blood cells as described in this chart.

Blood Type	Antigen on Red Blood Cells	Antibody in Plasma
A	A	Anti-B
B	B	Anti-A
AB	A, B	None
O	None	Anti-A and Anti-B

Blood type also indicates whether the person has an Rh factor, another type of antigen, on the red blood cells. This designation is attached to the ABO blood type as in A-positive or A-negative. It is customary to simply attach a positive or negative sign to A-negative and A-positive, which are compound words.

Use the precautions given in the warning box, because the synthetic products used in this exercise can be harmful to the skin.

> ⚠ Wear protective laboratory clothing, latex gloves, and goggles. If the chemicals touch the skin, eyes or mouth, wash immediately. If inhaled, seek fresh air.

1. Obtain 3 testing plates, each of which contains three depressions; vials of persons 1, 2, and 3 blood; vials of anti-A serum, anti-B serum, and anti-Rh serum.
2. Using a wax pencil, number the plates so you know which plate is for which person #1, #2, or #3. Look carefully at the plate and notice the wells are designated as A, B, or Rh.
3. Being sure to close the cap to each vial in turn,

   Add a drop of person #1 blood to all three wells of plate #1—close the cap.

   Add a drop of anti-A (blue) to the well designated A—close the cap.

   Add a drop of anti-B (yellow) to the well designated B—close the cap.

   Add a drop of anti-Rh (clear) to the well designated Rh—close the cap.
4. Stir the contents of each well with a mixing stick of the correct color. After a few minutes, examine the wells for **agglutination,** i.e., granular appearances that indicate the blood type. (Rh-positive takes the longest to react.) If a person is AB-positive, which wells would show agglutination? _____
5. Record the blood type results for each person here:

   Person #1 __O−__   Brown

   Person #2 __A−__   Smith

   Person #3 __AB−__   Green

Blood Flow

Blood must circulate to serve the body. The heart pumps the blood, which moves away from the heart in **arteries** and **arterioles** and returns to the heart in **venules** and **veins. Capillaries** connect arterioles to venules. Capillaries branch to form capillary beds that close when precapillary sphincters contract; in this way, capillaries direct the blood flow to regions that need it most (Fig. 8.6). When a capillary bed is closed, blood moves directly from the arteriole to the venule by way of the arteriovenous shunt.

Figure 8.6 Anatomy of blood vessels.
a. Photomicrograph of anatomy of veins. **b.** Blood flow. **c.** Capillary bed.

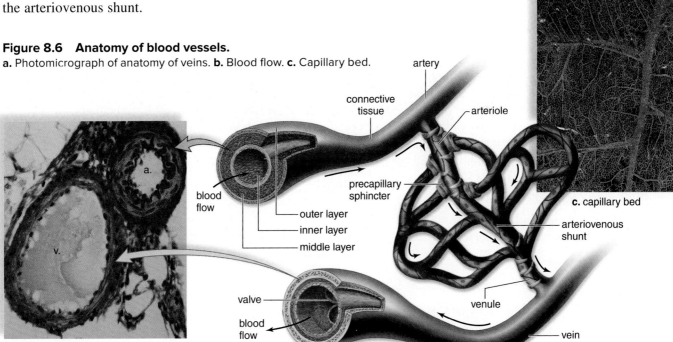

v = vein; a = artery
a. Photomicrograph

b. Blood flow

Observation: Blood Vessel Comparison

1. Obtain a microscope slide that shows an artery and a vein in cross section.
2. View the slide, under both low and high power; with the help of Figure 8.6a, determine which is the artery and which is the vein. Identify the **outer layer,** which contains many collagen and elastic fibers and often appears white in specimens. Identify the **middle layer,** the thickest layer, composed of

smooth muscle and elastic tissue. Does this layer appear thicker in arteries than in veins? _____

Identify the **inner layer,** a wavy lining of simple squamous epithelial cells called the endothelium. In veins, the endothelium forms valves that keep the blood moving toward the heart. Veins are larger and more "flabby" than arteries. Why might that be based on the structure of their walls?

_____

Experimental Procedure: Blood Flow

Observe blood flow through arterioles, capillaries, and venules, either in a videoclip, in the tail of a goldfish, or in the webbed skin between the toes of a frog. Observe the pulse and the swiftly moving blood in the arterioles. Contrast this with the more slowly moving blood that circulates in the opposite direction in the venules. Many criss-crossing capillaries are visible. Look for blood cells floating in the bloodstream.

Do you see any? _____

Conclusions: Blood Flow

- Which type of blood vessel (arteries or veins) has thicker walls? _____ Why is this advantageous? _____
- Which type of blood vessel has thinner walls? _____ Why is this advantageous?

_____

- Which type of blood vessel has valves? _____ Compare the direction of the flow of blood in the aorta to that in the inferior vena cava. Why is this advantageous for veins to have valves? _____

Path of Blood in Adult Humans

The right side of the heart pumps blood into the **pulmonary circuit**—to the lungs and back to the heart (Fig. 8.7). While the blood is in the lungs, it gives up carbon dioxide and gains oxygen. The left side of the heart, consisting of two chambers, pumps blood to the **systemic circuit**—throughout the whole body except the lungs. Blood in the systemic circuit gives up oxygen and gains carbon dioxide.

Figure 8.7 shows how to trace the path of blood in both the pulmonary and systemic circuits in adult humans. It will also assist you in learning the names of some of the major blood vessels.

Pulmonary and Systemic Circuits

1. Trace the path of blood in the pulmonary circuit from a chamber of the heart to the lungs, and then from the lungs to a chamber of the heart. Follow the arrows in Figure 8.7, and use the names of the blood vessels provided there. _____

_____

2. Trace the path of blood in the systemic circuit from the heart to the kidneys, and then from the kidneys to the heart. _____

Figure 8.7   Diagram of the human cardiovascular system.

In the pulmonary circuit, the pulmonary arteries take O_2-poor blood to the lungs, and the pulmonary veins return O_2-rich blood to the heart. In the systemic circuit, the aorta branches into the various arteries that go to all other parts of the body. After blood passes through arterioles, capillaries, and venules, it enters various veins and then the superior and inferior venae cavae, which return it to the heart.

3. Is it true that all arteries carry O_2-rich blood and all veins carry O_2-poor blood? _____

   Explain the color of the pulmonary blood vessels in Figure 8.7. _____

   _____

4. A portal system is a vein that begins in capillaries and ends in capillaries. For example, the hepatic portal vein begins in capillaries at the digestive tract and ends in capillaries at the liver. Trace the path of blood from the aorta to the inferior vena cava by way of the hepatic portal vein. _____

   _____

   _____

Names of Blood Vessels

Use Figures 8.7 and 8.8 to complete Table 8.3 by stating the name of the artery that takes blood to the body part and the name of the vein that takes blood away from the body part.

Figure 8.8   The major arteries and veins of the systemic circuit.
This illustration offers a more realistic representation of major blood vessels (arteries and veins) of the systemic circuit.

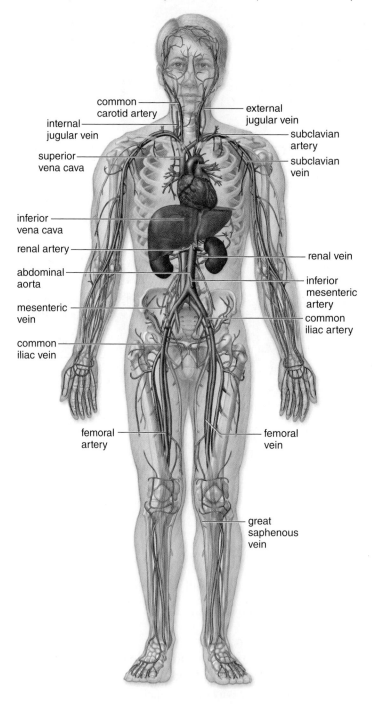

Table 8.3 Major Blood Vessels in the Systemic Circuit

Body Part	Artery	Vein
Heart		
Head		
Arms		
Kidney		
Legs		
Intestines		

Application for Daily Living

High Blood Pressure

Blood pressure is an indication of cardiovascular health because high blood pressure at rest is associated with clogged arteries that impede blood flow. Blood should flow freely to all parts of the body and especially to the heart because it has to pump the blood to keep it moving about the body. Unfortunately, the carotid arteries, which serve the heart, are the ones most apt to become clogged with fatty deposits called plaque.

Optimal blood pressure at rest is at or below 120/80. (These numbers are in mm Hg, a way to measure pressure, and systolic blood pressure is given first.) Prehypertension occurs when systolic pressures are between 120 and 139 and diastolic pressures are between 80 and 89 mm Hg. Hypertension, or high blood pressure, occurs when systolic pressure exceeds 140 and diastolic pressure exceeds 90, as in 150/95.

Age (over 55), male gender, and African-American ethnicity are all risk factors for developing plaque in the arteries. However, blood pressure can be improved through adequate exercise, good diet, loss of weight, not smoking, and reduced stress.

White _____ **1.** What type of blood cells are lymphocytes and monocytes?

_____ **2.** Nutrients exit and wastes enter which type of blood vessel?

Red _____ **3.** Which blood cells contain a respiratory pigment?

B atrium _____ **4.** Which chamber of the heart receives venous blood from the systemic circuit?

aortic valve _____ **5.** Identify the vessel that conducts blood from the left ventricle.

D ventricle _____ **6.** The pulmonary trunk leaves which chamber?

coronary _____ **7.** Identify the artery that nourishes the heart tissue.

D ventricle _____ **8.** Which heart chamber pumps blood to the body proper?

Systolic _____ **9.** Which is higher—systolic or diastolic pressure?

SA. Node _____ **10.** Identify the pacemaker region of the heart.

O2 poor _____ **11.** Does the pulmonary artery in adults carry O_2-rich or O_2-poor blood?

O Blood _____ **12.** Which type of blood shows no reaction to anti-A and anti-B serum.

Renal _____ **13.** Identify the artery that serves the kidney.

iliac (femoral) _____ **14.** Identify the arteries that take blood from the aorta to the legs.

arms _____ **15.** What part of the human body is served by a subclavian vessel?

Superior V.C. _____ **16.** Identify the large abdominal vein that runs alongside the aorta and enters the right atrium.

artery _____ **17.** Which type of blood vessel (artery or vein) has valves?

Thought Questions

18. What might be the significant result of nonfunctional or defective chordae tendineae?

19. During a heart attack, cardiac muscle cells are deprived of their blood supply, yet the atria are still receiving blood. Explain.

20. Evaluate the following statement: All arteries carry O_2-rich blood and all veins carry O_2-poor blood. Based on what you have learned in this laboratory, is this statement correct? Explain.

connect®

|BIOLOGY

Find even more resources to learn the chapter concepts including animations, tutorial videos, and interactive practice questions.

McGraw-Hill Access Science Website

An online encyclopedia of science and technology that provides information, including videos, that can enhance the laboratory experience.

www.accessscience.com

9

Chemical Aspects of Digestion

Learning Outcomes

Introduction
- Sequence the organs of the digestive tract from the mouth to the anus.
- State the contribution of each organ, if any, to the process of chemical digestion.

9.1 Protein Digestion by Pepsin
- Associate the enzyme pepsin with the ability of the stomach to digest protein.
- Explain why stomach contents are acidic and how a warm body temperature aids digestion.

Prelab Question: An acid pH plays what role in digestion by pepsin?

9.2 Fat Digestion by Pancreatic Lipase
- Associate the enzyme lipase with the ability of the small intestine to digest fat.
- Explain why the emulsification process assists the action of lipase.
- Explain why a change in pH indicates that fat digestion has occurred.
- Explain the relationship between time and enzyme activity.

Prelab Question: Sufficient time plays what role in digestion by any enzyme?

9.3 Starch Digestion by Pancreatic Amylase
- Associate the enzyme pancreatic amylase with the ability of the small intestine to digest starch.

Prelab question: A denatured enzyme cannot carry out digestion. Explain why not.

9.4 Requirements for Digestion
- Assuming a specific enzyme, list four factors that can affect the activity of all enzymes.
- Explain why the operative procedure that reduces the size of the stomach causes an individual to lose weight.

Prelab Question: Of the factors that can affect enzymatic action, which causes an enzyme to be unable to function to any degree?

Introduction

In Laboratory 7, you examined the organs of digestion in the fetal pig. Now we wish to further our knowledge of the digestive process by associating certain digestive enyzmes with particular organs, as shown in Figure 9.1. This laboratory will also give us an opportunity to study the action of enzymes, much as William Beaumont did when he removed food samples through a hole in the stomach wall of his patient, Alexis St. Martin. Every few hours, Beaumont would see how well the food had been digested.

In Laboratory 5 we learned that enzymes are very specific and usually participate in only one type of reaction. The active site of an enzyme has a shape that accommodates its substrate, and if an environmental factor such as a boiling temperature or a wrong pH alters this shape, the enzyme loses its ability to function well, if at all. We will have an opportunity to make these observations with controlled experiments. The box on the next page reviews what is meant by a controlled experiment.

> ⏱ **Planning Ahead** Be advised that protein digestion (page 107) requires 1½ hours and fat digestion (page 109) requires 1 hour. Also a boiling water bath is required for starch digestion (page 111).

Figure 9.1 Organs of the digestive tract (right) and accessory organs (left).

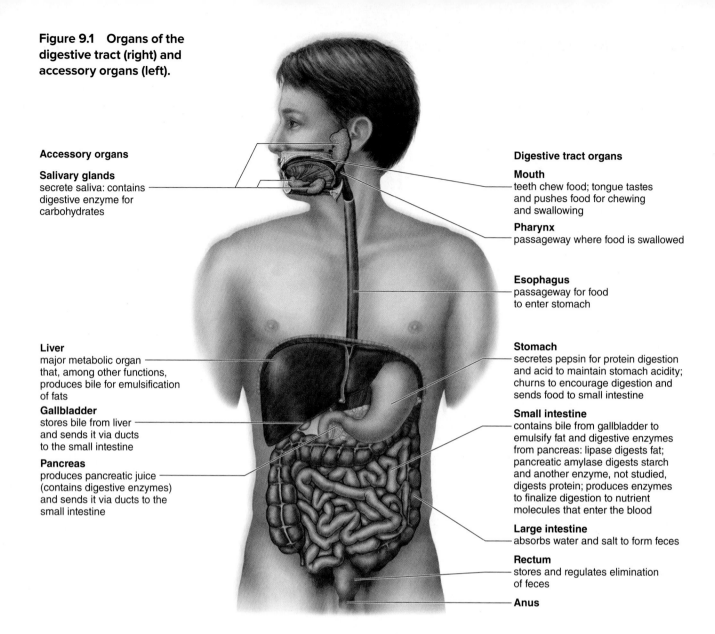

Accessory organs

Salivary glands
secrete saliva: contains digestive enzyme for carbohydrates

Liver
major metabolic organ that, among other functions, produces bile for emulsification of fats

Gallbladder
stores bile from liver and sends it via ducts to the small intestine

Pancreas
produces pancreatic juice (contains digestive enzymes) and sends it via ducts to the small intestine

Digestive tract organs

Mouth
teeth chew food; tongue tastes and pushes food for chewing and swallowing

Pharynx
passageway where food is swallowed

Esophagus
passageway for food to enter stomach

Stomach
secretes pepsin for protein digestion and acid to maintain stomach acidity; churns to encourage digestion and sends food to small intestine

Small intestine
contains bile from gallbladder to emulsify fat and digestive enzymes from pancreas: lipase digests fat; pancreatic amylase digests starch and another enzyme, not studied, digests protein; produces enzymes to finalize digestion to nutrient molecules that enter the blood

Large intestine
absorbs water and salt to form feces

Rectum
stores and regulates elimination of feces

Anus

What Is a Control?

The experiments in today's laboratory have both a positive control and a negative control, *which should be saved for comparison purposes until the experiment is complete*. The **positive control** goes through all the steps of the experiment and does contain the substance being tested. Therefore, positive results are expected. The **negative control** goes through all the steps of the experiment, except it does not contain the substance being tested. Therefore, negative results are expected.

For example, if a test tube contains glucose (the substance being tested) and Benedict's reagent (blue) is added, a red color develops upon heating. This test tube is the positive control; it tests positive for glucose. If a test tube does not contain glucose and Benedict's reagent is added, Benedict's is expected to remain blue. This test tube is the negative control; it tests negative for glucose.

What benefit is a positive control? Positive controls give you a standard by which to tell if the substance being tested is present (or acting properly) in an unknown sample. Negative controls ensure that the experiment is giving reliable results; after all, if a negative control should happen to give a positive result, then the entire experiment may be faulty and unreliable.

9.1 Protein Digestion by Pepsin

Certain foods, such as meat and egg whites, are rich in protein. Egg whites contain albumin, which is the protein used in this Experimental Procedure. Protein is digested by **pepsin** in the stomach (Fig. 9.2), a process described by the following reaction:

$$\text{protein} + \text{water} \xrightarrow{\text{pepsin (enzyme)}} \text{peptides}$$

The stomach has a very low pH. Does this indicate that pepsin works effectively in an acidic or a basic environment? _____ This is the pH that allows the enzyme to maintain its normal shape so that it will combine with the substrate. A warm temperature causes molecules to move about more rapidly and increases the encounters between enzyme and substrate. Therefore you would hypothesize that the yield from this enzymatic reaction will be higher if the pH is _____ and the temperature is _____ (body temperature 37°C).

Test for Protein Digestion

Biuret reagent is used to test for protein digestion. If digestion has not occurred, biuret reagent turns purple, indicating that protein is present. If digestion has occurred, biuret reagent turns pinkish-purple, indicating that peptides are present.

> ⚠️ **Biuret reagent** is highly corrosive. Exercise care in using this chemical. If any should spill on your skin, wash the area with mild soap and water. Follow your instructor's directions for its disposal.

Experimental Procedure: Protein Digestion

1. Label four clean test tubes (1 to 4). Using the designated graduated pipet, add 2 ml of the albumin solution to all tubes. Albumin is a protein.
2. Add 2 ml of the pepsin solution to tubes 1 to 3, as listed in Table 9.1.
3. Add 2 ml of 0.2% HCl to tubes 1 and 2. HCl simulates the acidic conditions of the stomach.
4. Add 2 ml of water to tube 3 and 4 ml of water to tube 4, as listed in Table 9.1.
5. Swirl to mix the tubes. Tube 2 remains at room temperature, but the other three are incubated for 1½ hours. Record the temperature for each tube in Table 9.1.
6. Remove the tubes from the incubator and place all four tubes in a tube rack. Add 2 ml of biuret reagent to all tubes and observe. Record your results in Table 9.1 as + or − to indicate digestion or no digestion.

Figure 9.2 Digestion of protein.
Pepsin, produced by the gastric glands of the stomach, helps digest protein.

Table 9.1 Protein Digestion by Pepsin

Tube	Contents	Temperature	Digestion (+ or −)	Explanation
1	Albumin Pepsin HCl Biuret reagent			
2	Albumin Pepsin HCl Biuret reagent			
3	Albumin Pepsin Water Biuret reagent			
4	Albumin Water Biuret reagent			

Conclusions: Protein Digestion

- Explain your results in Table 9.1 by reasoning why digestion did or did not occur. To be complete, consider all the requirements for an enzymatic reaction as listed in Table 9.4. Now show here that tube 1 met all the requirements for digestion:

  Pepsin is the correct _____.

  Albumin is the correct _____.

  37°C is the optimum _____.

  HCl provides the optimum _____.

  1½ hours provides _____ for the reaction to occur.

- Review "What Is a Control?" on page 106. Which tube was the negative control? _____

  Explain why it was the negative control. _____

  _____

- If this control tube had given a positive result for protein digestion, what could you conclude about

  this experiment? _____

  _____

9.2 Fat Digestion by Pancreatic Lipase

Lipids include fats (e.g., butterfat) and oils (e.g., sunflower, corn, olive, and canola). Lipids are digested by **pancreatic lipase** in the small intestine (Fig. 9.3).

Figure 9.3 Emulsification and digestion of fat.
Bile from the liver (stored in the gallbladder) enters the small intestine, where lipase in pancreatic juice from the pancreas digests fat.

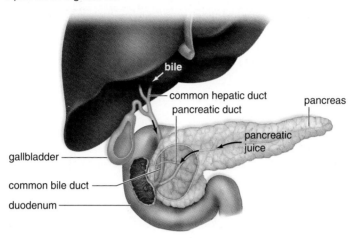

The following two reactions describe fat digestion:

1.
$$\text{fat} \xrightarrow{\text{bile (emulsifier)}} \text{fat droplets}$$

2.
$$\text{fat droplets + water} \xrightarrow{\text{lipase (enzyme)}} \text{glycerol + fatty acids}$$

With regard to the first step, consider that fat is not soluble in water; yet, lipase makes use of water when it digests fat. Therefore, bile is needed to emulsify fat—cause it to break up into fat droplets that disperse in water. The reason for dispersal is that bile contains molecules with two ends. One end is soluble in fat, and the other end is soluble in water. Bile can emulsify fat because of this.

With regard to the second step, would the pH of the solution be lower before or after the enzymatic reaction? (*Hint:* Remember that an acid decreases pH and a base increases pH.) _____

_____

Test for Fat Digestion

In the test for fat digestion, you will be using a pH indicator, which changes color as the solution in the test tube goes from basic conditions to acidic conditions. Phenol red is a pH indicator that is red in basic solutions and yellow in acidic solutions.

Experimental Procedure: Fat Digestion

1. Label three clean test tubes (1 to 3). Using the designated graduated pipet, add 1 ml of vegetable oil to all tubes.
2. Add 2 ml of phenol red solution to each tube. What role does phenol red play? _____
3. Add 2 ml of pancreatic lipase (pancreatin) to tubes 1 and 2 and 2 ml of water to tube 3, as listed in
   Table 9.2. What role does lipase play? _____
4. Add a pinch of bile salts to tube 1.
5. Record the initial color of all tubes in Table 9.2.
6. Incubate all three tubes at 37°C and check every 20 minutes.
7. Record any color change and how long it took to see this color change in Table 9.2.

Table 9.2 Fat Digestion by Pancreatic Lipase

Tube	Contents	Color		Time Taken	Explanation
		Initial	Final		
1	Vegetable oil Phenol red Pancreatin Bile salts				
2	Vegetable oil Phenol red Pancreatin				
3	Vegetable oil Phenol red Water				

Conclusions: Fat Digestion

- Explain your results in Table 9.2 by reasoning why digestion did or did not occur.
- What role did bile salts play in this experiment? _____

  _____
- What role did phenol red play in this experiment? _____
- Review "What Is a Control?" on page 106. Which test tube in this experiment could be considered a negative control? _____

  _____

9.3 Starch Digestion by Pancreatic Amylase

Starch is present in bakery products and in potatoes, rice, and corn. Starch is digested by **pancreatic amylase** in the small intestine, a process described by the following reaction:

$$\text{starch} + \text{water} \xrightarrow{\text{amylase (enzyme)}} \text{maltose}$$

1. If digestion *does not* occur, which will be present—starch or maltose? _____

2. If digestion *does* occur, which will be present—starch or maltose? _____

Tests for Starch Digestion

You will be using two tests for starch digestion:

1. If digestion has not taken place, the iodine test for starch will be positive (+). If digestion has occurred, the iodine test for starch will be negative (−).
2. If digestion of starch has taken place, the Benedict's test for sugar (maltose) will be positive (+). If digestion has not taken place, the Benedict's test for sugar will be negative (−).

> ⚠ **Benedict's reagent** is highly corrosive. Use protective eyewear when performing this experiment. Exercise care in using this chemical. If any should spill on your skin, wash the area with mild soap and water. Follow your instructor's directions for disposal of this chemical.

To test for sugar, add five drops of Benedict's reagent. Place the tube in a boiling water bath for a few minutes, and note any color changes (see Table 9.3 on page 112). Boiling the test tube is necessary for the Benedict's reagent to react.

Experimental Procedure: Starch Digestion

1. Label six clean test tubes (1 to 6).
2. Using a transfer pipet, add 1 ml of pancreatic-amylase solution to tubes 1 to 4 and 1 ml of water to tubes 5 and 6.
3. Test tubes 1 and 2 immediately.

   Tube 1 Shake the starch solution and add 1 ml of starch solution. Immediately add five drops of iodine to test for starch. Put this tube in a test tube rack and record your results in Table 9.3.

   Tube 2 Shake the starch solution and add 1 ml of starch solution. Immediately add five drops of Benedict's reagent, and place the tube in a boiling water bath to test for sugar. Put this tube in the test tube rack and record your results in Table 9.3.

4. Shake the starch suspension and add 1 ml of starch suspension to tubes 3 to 6. Allow the tubes to stand for 30 minutes.

   Tubes 3 and 5 After 30 minutes, test for starch using the iodine test. Place these tubes in the test tube rack and record your results in Table 9.3.

   Tubes 4 and 6 After 30 minutes, test for sugar using the Benedict's test. Place these tubes in the test tube rack and record your results in Table 9.3.

5. Examine all your tubes in the test rube rack and decide whether digestion occurred (+) or did not occur (−). Complete Table 9.3.

Table 9.3 Starch Digestion by Amylase

Tube	Contents	Time*	Type of Test	Test Resuts (+ or −)	Digestion (+ or −)
1	Pancreatic amylase Starch	0	Iodine	+	
2	Pancreatic amylase Starch				
3	Pancreatic amylase Starch				
4	Pancreatic amylase Starch				
5	Water Starch				
6	Water Starch				

* Enter either 0 for immediately or T for after 30 minutes.

Conclusions: Starch Digestion

- Considering tubes 1 and 2, this experimental procedure showed that _____ must pass for digestion to occur.
- Considering tubes 5 and 6, this experimental procedure showed that an active _____ must be present for digestion to occur.
- Why would you not recommend doing the test for starch and the test for sugar on the same tube? _____
  _____
- Which test tubes served as a negative control for this experiment? _____
  Explain your answer. _____

Absorption of Sugars and Other Nutrients

Figure 9.4 shows that the folded lining of the small intestine has many fingerlike projections called villi. The small intestine not only digests food; it also absorbs the products of digestion, such as sugars from carbohydrate digestion, amino acids from protein digestion, and glycerol and fatty acids from fat digestion at the villi.

Figure 9.4 Anatomy of the small intestine.
Nutrients enter the bloodstream across the much-convoluted walls of the small intestine.

9.4 Requirements for Digestion

Explain in Table 9.4 how each of the requirements listed influences effective digestion.

Table 9.4 Requirements for Digestion	
Requirement	**Explanation**
Specific enzyme	
Specific substrate	
Warm temperature	
Specific pH	
Time	
Fat emulsifier	

To lose weight, some obese individuals undergo an operation in which (1) the stomach is reduced to the size of a golf ball, and (2) food bypasses the duodenum (first 2 feet) of the intestine. Answer these questions to explain how this operation would affect the requirements for digestion.

1. How is the amount of substrate reduced? _____

2. How is the amount of digestive enzymes reduced? _____

3. How is time reduced? _____

4. What makes the pH of the small intestine higher than before? _____

5. How is fat emulsification reduced? _____

6. How does surgery to reduce obesity sometimes result in malnutrition? _____

———————————— 1. When iodine (IKI) solution turns blue-black, what substance is present?

———————————— 2. What color is Benedict's reagent originally?

———————————— 3. What happens to an enzyme when it is boiled?

———————————— 4. Saliva and pancreatic juice contain what type enzyme to digest starch?

———————————— 5. As oil is digested, why does the solution turn from red to yellow, indicating an acid pH?

———————————— 6. What temperature promotes enzymatic action?

———————————— 7. What do you call a sample that goes through all the steps of an experiment but lacks the factor being tested?

———————————— 8. What role do bile salts play in the digestion of fat?

———————————— 9. What color does biuret reagent turn when peptides are present?

———————————— 10. Is the optimal pH for pepsin acidic or basic?

———————————— 11. Why would you predict that pepsin would not digest starch?

———————————— 12. In addition to pepsin and water, what is needed to digest protein?

———————————— 13. Name the enzyme responsible for the hydrolysis of starch.

Thought Questions

14. Which of the following two combinations is most likely to result in digestion?

    a. Pepsin, protein, water, body temperature

    b. Pepsin, protein, hydrochloric acid (HCl), body temperature

    Explain.

15. Which of the following two combinations is most likely to result in digestion?

    a. Amylase, starch, water, body temperature, testing immediately

    b. Amylase, starch, water, body temperature, waiting 30 minutes

    Explain.

|BIOLOGY

Find even more resources to learn the chapter concepts including animations, tutorial videos, and interactive practice questions.

McGraw-Hill Access Science Website

An online encyclopedia of science and technology that provides information, including videos, that can enhance the laboratory experience.

www.accessscience.com

10

Energy Requirements and Ideal Weight

Learning Outcomes

10.1 Average Daily Energy Intake
- Keep a food diary and use it to calculate your average daily energy intake.

Prelab Question: What unit of measurement is used for daily energy intake?

10.2 Average Daily Energy Requirement
- Keep a physical activity diary and use it to calculate average daily energy required for physical activity.
- Calculate the daily energy required for your basal metabolism.
- Calculate the daily energy required for your specific dynamic action.
- Calculate your total average daily energy requirement.

Prelab Question: Daily energy requirement has what three components?

10.3 Comparison of Average Daily Energy Intake and Average Daily Energy Requirement
- Predict whether weight gain or weight loss will occur after energy intake is compared to energy required.

Prelab Question: What two energy measurements determine whether you gain weight or lose weight?

10.4 Ideal Weight
- Calculate ideal weight based on body mass index (BMI).
- Calculate ideal weight based on body composition.
- Make a recommendation regarding daily energy intake and daily energy required for achieving or maintaining ideal weight.

Prelab Question: What are two methods for calculating ideal weight?

Application for Daily Living: Losing Weight

Introduction

This laboratory helps you determine your average daily energy (kcal) requirements, your average daily energy (kcal) intake, and your ideal weight. In addition to the energy content of food, other nutritional aspects are also important. Minerals and vitamins, along with specific types of biological molecules, contribute to a healthy diet. So, while the focus of this lab is on energy in our food, there is more to diet than just calories.

Before you come to the laboratory, it is necessary for you to keep two daily diaries for three days. The physical activity diary is a record of the energy you expend for physical activity. Physical activity is only a portion of your daily energy requirement. You also need energy for **basal metabolism** (energy needed when the body is resting) and **specific dynamic action** (energy needed to process food), also known as thermal effect of food.

The food diary is a record of the energy you take in as food. If the energy you expend and the energy you take in are in balance, you will neither gain nor lose weight. If your ideal weight is different from your present weight, you will be able to determine whether to increase or decrease your physical activity and food intake to attain an ideal weight.

10.1 Average Daily Energy Intake

To complete this laboratory, you must calculate an average daily energy intake. The first method given is preferred because it makes use of your actual intake. However, when appropriate, a hypothetical method is also provided.

Personal Diet

Follow these directions.

1. Keep a written food diary of your own design for three days. Your diary should list all food and beverages consumed throughout an entire three-day period, including snacks and alcoholic beverages.

2. Make sure your food diary contains a column for recording the energy in kcal (kilocalories)* of the food you eat. Table 10.1 may be helpful to you in this regard. Also, food packages often contain kcal information, as do various calorie books and computer programs. For example, after eating two slices of bread, your diary might look like this:

Day of Week _____			
Type of Food	**Portion Size**	**kcal**	**Total kcal**
Bread	Two slices	70/one slice	140

Continue in this manner until you have listed all the foods you eat in one day. Do the same for two more days.

3. Add up your total kcal intake for each day, and record the information here:

   Day 1 energy intake _____ kcal

   Day 2 energy intake _____ kcal

   Day 3 energy intake _____ kcal

   Total _____ kcal

4. Divide the total by 3 to calculate the average daily energy intake.

> Av. energy intake/day = _____ kcal

---

*The amount of heat required to raise 1 kg (kilogram) of water 1°C.

Table 10.1 Nutrient and Energy Content of Foods*

Breads

1 slice of bread or any of the following:

¾ cup ready-to-eat cereal

⅓ cup corn

1 small potato

(1 bread = 15 g carbohydrate, 2 g protein, and 70 kcal)

Milk

1 cup skim milk or: (2% milk—add 5 g fat)

1 cup skim-milk yogurt, plain   (Whole milk—add 8 g fat)

1 cup buttermilk

½ cup evaporated skim milk or milk dessert

(1 milk = 12 g carbohydrate, 8 g protein, and 80 kcal)

Vegetables

½ cup greens

½ cup carrots

½ cup beets

(1 vegetable = 5 g carbohydrate, 1 g protein, and 25 kcal)

Fruits

½ small banana or

1 small apple

½ cup orange juice or ½ grapefruit

(1 fruit = 10 g carbohydrate and 40 kcal)

Meats (lean)

1 oz. lean meat or:

1 oz. chicken meat without the skin

1 oz. any fish

¼ cup canned tuna or 1 oz. low-fat cheese

(1 oz. low-fat meat = 7 g protein, 3 g fat, and 55 kcal)

Meats (medium fat)

1 oz. pork loin

1 egg

¼ cup creamed cottage cheese

(1 medium-fat meat = 7 g protein,

5½ g fat, and about 80 kcal)

Meats (high fat)

1 oz. high-fat meat is like:

1 oz. country-style ham

1 oz. cheddar cheese

A small hot dog (frankfurter)

(1 high-fat meat = 7 g protein, 8 g fat, and
100 kcal)

Peanut butter

Peanut butter is like a meat in terms of its protein
content but is very high in fat. It is estimated as
(2 tbsp peanut butter = 7 g protein, 15½ g fat, and
about 170 kcal)

Fats

1 tsp butter or margarine

1 tsp any oil

1 tbsp salad dressing

1 strip crisp bacon

5 small olives

10 whole Virginia peanuts

(1 fat = 5 g fat and 45 kcal)

Legumes (beans and peas)

Legumes are like meats because they are rich in
protein and iron but are lower in fat than meat.
They contain much starch. They can be treated as
(½ cup legumes = 15 g carbohydrate, 9 g protein,
3 g fat, and 125 kcal)

Miscellaneous Foods								
	g (grams)			*kcal*		*g (grams)*		*kcal*
	Protein	Fat	Carbohydrate			Protein	Fat	Carbohydrate	
Ice cream (1 cup)	5	14	32	274	Beer (1 can)	1	0	14	60
Cake (1 piece)	3	1	32	149	Soft drink	0	0	37	148
Doughnuts (1)	3	11	22	199	Soup (1 cup)	2	3	15	95
Pie	3	15	51	351	Coffee and tea	0	0	0	0
Caramel candy (1 oz.)	1	3	22	119					

*Use this table for the Food Diary mentioned on p. 116.

Hypothetical Diet

Suppose you have decided to eat your meals at the fast-food restaurants listed in the table.

1. Circle the restaurant and the items you have chosen for breakfast, lunch, and dinner. Choose a different restaurant for each meal.
2. Record the number of kcal for the items chosen.
3. Add the number of kcal.
4. This will be your average energy intake/day.

Fast-Food Menus and kcal

Restaurant/Menu	kcal	Restaurant/Menu	kcal
Taco Bell (www.tacobell.com)		**McDonald's (www.mcdonalds.com)**	
Steak Baja chalupa	390	Egg McMuffin	300
Soft beef taco supreme	240	Sausage McMuffin w/egg	450
Chicken Baja gordita	320	Ham and egg cheese bagel	550
Spicy chicken soft taco	170	Hash browns	130
Nachos Bellgrande	760	Hotcakes (w/margarine & syrup)	350
Soft drink (small)	150	Breakfast burrito	290
Subway (www.subway.com)		6 pc. chicken nugget	280
Meatball marinara	560	Barbeque sauce	50
6" roasted chicken breast	310	Hamburger	250
6" roast beef	290	Quarter pounder w/cheese	510
Soft drink (small)	150	Big Mac	540
Burger King (www.burgerking.com)		Filet-O-Fish	380
Croissan'wich w/sausage, egg & cheese	520	Chicken McGrill w/mayo	420
Croissan'wich w/egg & cheese	320	Med. french fries	380
Egg'wich™ w/Canadian bacon & egg	380	Large french fries	500
French toast sticks	390	Hot fudge sundae	340
Lg. hash brown rounds	390	Vanilla milk shake	570
Double Whopper w/Cheese and mayo	1150	Soft drink (small)	150
Whopper w/cheese and mayo	760		
BK big fish	710	**Papa John's (www.papajohns.com)**	
King-sized fries w/salt	600	Cheese pizza	283
Med.-sized fries w/salt	360	All-meat pizza	390
King-sized onion rings	550	Garden special pizza	280
Med.-sized onion rings	320	The works pizza	342
Dutch apple pie	340	Cheese sticks	180
Hershey's sundae pie	310	Breadsticks	140
Soft drink (small)	150	Garlic sauce	235
Vanilla milk shake (medium)	720	Pizza sauce	25
		Nacho cheese sauce	60
		Soft drink (small)	150

Av. energy intake/day = _____ kcal

10.2 Average Daily Energy Requirement

As stated at the beginning of this laboratory, the body needs energy for three purposes: (1) energy for physical activity, (2) energy to support basal metabolism, and (3) energy for the specific dynamic action of processing food. We will calculate the daily energy requirement for each of these.

Calculating Average Daily Energy Required for Physical Activity

You will need to know your average daily energy requirement for physical activity before coming to this laboratory.

1. Keep a physical activity diary for three days. Make three copies of Table 10.2, one for each day.
2. Use Scale 3 (left and right sides) in Figure 10.1, p. 121, to convert your body weight to kg (kilograms). Place this value on the appropriate line in your physical activity diary (Table 10.2).
3. Consult Table 10.3 and fill in column 4 of your physical activity diary.
4. Fill in column 5 of your physical activity diary by using a calculator and this formula:

$$\text{time spent} \times \text{energy cost} \times \text{body weight} = \text{total energy expended}$$
$$\text{(min)} \qquad\qquad\qquad\qquad \text{(kg)} \qquad\qquad \text{kcal}$$

Therefore, if 100 minutes were spent doing word processing, and the body weight is 77.3 kg, the kcal expended for typing would be

$$100 \text{ min} \times 0.01 \text{ kcal/kg min} \times 77.3 \text{ kg} = 77.3 \text{ kcal}$$

5. Total the number of kcal expended for each of the three days, and divide this total by 3 to get the average (av.) daily energy required for physical activity:

Energy for physical activity (day 1) _____ kcal (see p. 120)

Energy for physical activity (day 2) _____ kcal

Energy for physical activity (day 3) _____ kcal

Total energy for physical activity/three days _____ kcal

Av. energy required for physical activity/day _____ kcal

A completed food diary and physical activity diary are necessary before you begin the laboratory.

Table 10.2 Physical Activity Diary for One Day

Time of Day	Activity	Time Spent (min)	Factor (kcal/kg min)	Weight (kg)	Total Energy Expended (kcal)

Total energy expended _____

Table 10.3 Energy Cost for Activities (Exclusive of Basal Metabolism and Specific Dynamic Action)

Type of Activity	Energy Cost (kcal/kg min)	Type of Activity	Energy Cost (kcal/kg min)
Sitting or standing still	0.010	Heavy exercise	0.070
Studying		Fast dancing	
Writing		Walking uphill	
Word processing		Jogging	
TV watching		Fast swimming	
Eating		Severe exercise	0.110
Very light activity	0.020	Tennis	
Driving car		Racquetball	
Walking slowly		Running	
Light exercise	0.025	Aerobic dancing	
Light housework		Soccer	
Walking at moderate speed		Very severe exercise	0.140
Carrying books or packages		Wrestling	
Moderate exercise	0.040	Boxing	
Fast walking		Racing	
Slow dancing		Rowing	
Slow bicycling		Full-court basketball	
Golf			

Calculating BMR/Day

The **basal metabolic rate** (**BMR**) is the rate at which kcal are spent for basal metabolism. Basal metabolism is the minimum amount of energy the body needs at rest in the fasting state. The beating of the heart, breathing, maintaining body temperature, and sending nerve impulses are some of the activities that maintain life. BMR varies according to a person's body surface area, age, and gender. On page 122, you will multiply your body surface area by a basal metabolic rate constant to arrive at your BMR/hour. This hourly rate multiplied by 24 hr/day will give you the BMR/day.

Figure 10.1 Method to estimate body surface area from height and weight.
A straight line is drawn from the subject's height (Scale 1) to the subject's weight (Scale 3). The point at which the line intersects Scale 2 is the subject's body surface area in m^2 (meters squared).

Body Surface Area

1. Use a scale to determine your weight and a measuring device (e.g., measuring tape) to determine your height. It is assumed that you are fully clothed and wearing shoes with a 1-inch heel. **Be honest about your weight.** There is no need to tell anyone else.

2. Consult Figure 10.1 and with a ruler, draw a straight line from your height to your weight. The point where that line crosses the middle column shows your surface area in m^2 (squared meters). For example, a person who is 6 ft tall and weighs 170 lb has a body surface area of 1.99 m^2.

   What is your body surface area? _____

BMR/Day

1. Consult Table 10.4 to find the BMR constant for your age and sex. Multiply your surface area by this factor to calculate your BMR/hr. For example, a 17-year-old male has a BMR constant of 41.5 kcal/m^2 hr. If his surface area is 1.99 m^2, his BMR is 1.99 m^2 × 41.5 kcal/m^2 = 82.6 kcal/hr.

   What is your BMR/hr? _____

2. Multiply your BMR/hr by 24 to obtain the total number of kcal you need for BMR/day. For example, if the BMR is 82.6 kcal/hr, then the daily BMR rate is 24 hr × 82.6 kcal/hr = 1,982 kcal/day.

   What is your BMR/day? _____

Table 10.4 Basal Metabolic Rate Constants

Age	BMR (kcal/m^2 hr) Males	BMR (kcal/m^2 hr) Females	Age	BMR (kcal/m^2 hr) Males	BMR (kcal/m^2 hr) Females
10	47.7	44.9	29	37.7	35.0
11	46.5	43.5	30	37.6	35.0
12	45.3	42.0	31	37.4	35.0
13	44.5	40.5	32	37.2	34.9
14	43.8	39.2	33	37.1	34.9
15	42.9	38.3	34	37.0	34.9
16	42.0	37.2	35	36.9	34.8
17	41.5	36.4	36	36.8	34.7
18	40.8	35.8	37	36.7	34.6
19	40.5	35.4	38	36.7	34.5
20	39.9	35.3	39	36.6	34.4
21	39.5	35.2	40–44	36.4	34.1
22	39.2	35.2	45–49	36.2	33.8
23	39.0	35.2	50–54	35.8	33.1
24	38.7	35.1	55–59	35.1	32.8
25	38.4	35.1	60–64	34.5	32.0
26	38.2	35.0	65–69	33.5	31.6
27	38.0	35.0	70–74	32.7	31.1
28	37.8	35.0	75+	31.8	

Calculating Energy Required for SDA

The **specific dynamic action** (**SDA**) refers to the amount of energy needed to process food. For example, muscles that move food along the digestive tract and glands that make digestive juices use up energy. To calculate the amount of average energy you require for daily SDA, (1) add the average energy required for daily physical activity and the energy for daily BMR. (2) Multiply the total by 10% to obtain an estimated average daily SDA. For example, if 1,044 kcal/day is required for physical activity and 1,882 kcal/day is required for BMR, then an average SDA = 293 kcal/day.

What is your **average SDA/day?** _____ kcal

Calculating Average Daily Energy Requirement

Total the amounts you have calculated for average daily physical activity, daily BMR, and average daily SDA. This is your average (av.) energy requirement/day.

Av. physical activity/day _____ kcal (see p. 119)

BMR/day _____ kcal (see p. 122)

Av. SDA/day _____ kcal (see p. 123, above)

Av. energy requirement/day = _____ kcal

10.3 Comparison of Average Daily Energy Intake and Average Daily Energy Requirement

Figure 10.2 illustrates that if the average daily energy intake is the same as the average daily energy requirement, weight is likely to remain the same.

1. Compare your average daily energy intake (p. 116 or p. 118) to your average daily energy

   requirement: _____

2. Our methods of calculation are only approximate. If your two figures are within 20% of each other, you will most likely neither lose nor gain weight. If the two figures are not within 20% of each other, are you apt to

   lose weight or gain weight? _____

   Explain. _____

   _____

Figure 10.2   Comparison of average daily energy intake and average daily energy requirement.
If average daily energy intake is the same as the average daily energy requirement, the person's weight stays the same.

10.4 Ideal Weight

Three methods of determining your ideal weight are explained next. Your instructor will choose which one you are to use.

Ideal Weight Based on Height and Weight Tables

Tables 10.5 and 10.6 are constructed by life insurance companies from data on thousands of people, at an age when bone and muscle growth is complete. Such tables are not highly accurate. Determine your ideal weight

Table 10.5	Weight and Height Table for Males			
Height		Small Frame	Medium Frame	Large Frame
(ft)	(in)	(lb)	(lb)	(lb)
5	2	128–134	131–141	138–150
5	3	130–136	133–143	140–153
5	4	132–138	135–145	142–156
5	5	134–140	137–148	144–160
5	6	136–142	139–151	146–164
5	7	138–145	142–154	149–168
5	8	140–148	145–157	152–172
5	9	142–151	148–160	155–176
5	10	144–154	151–163	158–180
5	11	146–157	154–166	161–184
6	0	149–160	157–170	164–188
6	1	152–164	160–174	168–192
6	2	155–168	164–178	172–197
6	3	158–172	167–182	176–202
6	4	162–176	171–187	181–207

Table 10.6	Weight and Height Table for Females			
Height		Small Frame	Medium Frame	Large Frame
(ft)	(in)	(lb)	(lb)	(lb)
4	10	102–111	109–121	118–131
4	11	103–113	111–123	120–134
5	0	104–115	113–126	122–137
5	1	106–118	115–129	125–140
5	2	108–121	118–132	128–143
5	3	111–124	121–135	131–147
5	4	114–127	124–138	134–151
5	5	117–130	127–141	137–155
5	6	120–133	130–144	140–159
5	7	123–136	133–147	143–163
5	8	126–139	136–150	146–167
5	9	129–142	139–153	149–170
5	10	132–145	142–156	152–173
5	11	135–148	145–159	155–176
6	0	138–151	148–162	158–179

Weight at ages 25–59 based on lowest mortality. Weight according to frame (indoor clothing weighing 3 lb is allowed for women; 5 lb for men), shoes with 1-inch heels.

range. A person more than 10% over the weight in the tables is overweight; a person 20% over is obese. Similarly, a person 10% below the weight in the table is underweight. What is your ideal weight based on these height and weight tables?

Ideal weight _____ lb

Ideal Weight Based on Body Mass Index (BMI)

Your **body mass index** (**BMI**) is a measure of your weight relative to your height (Table 10.7). A BMI of 19–25 is considered ideal. Below 19 a person is considered underweight, and above 25 a person is considered overweight. An obese person has a BMI of 30 or greater.

Table 10.7 Body Mass Index

Height (Inches)	19	20	21	22	23	24	25	26	27	28	29	30	31	32	33	34	35
				Ideal				Body Weight (Pounds) Not Ideal									
58	91	96	100	105	110	115	119	124	129	134	138	143	148	153	158	162	167
59	94	99	104	109	114	119	124	128	133	138	143	148	153	158	163	168	173
60	97	102	107	112	118	123	128	133	138	143	148	153	158	163	168	174	179
61	100	106	111	116	122	127	132	137	143	148	153	158	164	169	174	180	185
62	104	109	115	120	126	131	136	142	147	153	158	164	169	175	180	186	191
63	107	113	118	124	130	135	141	146	152	158	163	169	175	180	186	191	197
64	110	116	122	128	134	140	145	151	157	163	169	174	180	186	192	197	204
65	114	120	126	132	138	144	150	156	162	168	174	180	186	192	198	204	210
66	118	124	130	136	142	148	155	161	167	173	179	186	192	198	204	210	216
67	121	127	134	140	146	153	159	166	172	178	185	191	198	204	211	217	223
68	125	131	138	144	151	158	164	171	177	184	190	197	203	210	216	223	230
69	128	135	142	149	155	162	169	176	182	189	196	203	209	216	223	230	236
70	132	139	146	153	160	167	174	181	188	195	202	209	216	222	229	236	243
71	136	143	150	157	165	172	179	186	193	200	208	215	222	229	236	243	250
72	140	147	154	162	169	177	184	191	199	206	213	221	228	235	242	250	258
73	144	151	159	166	174	182	189	197	204	212	219	227	235	242	250	257	265
74	148	155	163	171	179	186	194	202	210	218	225	233	241	249	256	264	272
75	152	160	168	176	184	192	200	208	216	224	232	240	248	256	264	272	279
76	156	164	172	180	189	197	205	213	221	230	238	246	254	263	271	279	287

Procedure to Determine Ideal Weight Based on Body Mass Index (BMI)

1. Consult Table 10.7 and find your height in inches.

2. Go across that row to your weight.

3. Read the number at the top of that column. What is your BMI? _____

4. Is your BMI within the normal range (19–25)? _____ What is your ideal weight range based on BMI? _____

Comparison

How does this weight range compare to your desirable weight range from Table 10.5 or Table 10.6?

Ideal Weight Based on Body Composition

Body composition refers to the lean body weight plus the body fat. Ideal weight based on body composition is preferred because then you know how much of your weight is due to fat. In adult males, lean body weight should equal about 84% of weight, and body fat should be only 16% of weight. In adult females, lean body weight should equal about 77% of weight, and body fat should be only 23% of weight. Well-conditioned male athletes, such as marathon runners and swimmers, usually have 10–12% body fat, whereas football players can have as high as 19–20% body fat. At no time should the percentage of body fat drop below 5% in males and 10% in females.

To calculate **lean body weight** (**LBW**), it is necessary to subtract the amount of your body fat from your present weight. The amount of body fat is to be determined by taking skin-fold measurements, because the fat attached to the skin is roughly proportional to total body fat. The more areas of the body measured, the more accurate the estimate of body fat will be. In this laboratory, we will take only those measurements that permit you to remain fully clothed.

Procedure to Determine Lean Body Weight

1. Work in pairs. Obtain calipers designed to measure skin folds. All measurements should be done on the right side in mm (millimeters). Firmly grasp the fold of skin between the left thumb and the other four fingers and lift. Pinch and lift the fold several times to make sure no musculature is grasped. Hold the skin fold firmly, and place the contact side of the calipers 1/2 inch below the thumb and fingers; do not let go of the fold. Close the calipers on the skin fold, and do the measurements noted in items 2–6 following. Take the reading to the nearest 1/2 mm. Release the grip on the caliper and release the fold. To make sure the reading is accurate, repeat the measurement a few times. If the second measurement is within 1–2 mm of the first, it is reliable. Record two readings.
2. **Triceps (females only).** The triceps skin fold is measured on the back of the upper arm, halfway between the elbow and the tip of the shoulder, while the arm is hanging loosely at the subject's side. Grasp the skin fold parallel to the long axis of the arm and measure as described in step 1. Record in #7 below. Record two readings.
3. **Ilium—hip (males and females).** Fold the skin diagonally just above the top of the hip bone on an imaginary line that would divide the body into front and back halves. Measure and record as before. Record two readings.
4. **Abdomen (males and females).** Fold the skin vertically 1 inch to the right of the navel. Measure and record as before. Record two readings.
5. **Chest (males only).** Fold the skin diagonally, midway between the nipple and the armpit. Measure and record as before. Record two readings.
6. **Axilla—side (males only).** Fold the skin vertically at the level of nipple on an imaginary line that would divide the body into front and back halves. Measure and record as before. Record two readings.
7. Total any four of your skin-fold measurements.

	Females	**Males**
Triceps	_____ _____ mm	
Ilium	_____ _____ mm	_____ _____ mm
Abdomen	_____ _____ mm	_____ _____ mm
Chest		_____ _____ mm
Axilla		_____ _____ mm
Total of any four skin-fold measurements	_____ mm	_____ _____ mm

8. Consult Table 10.8 (males) and Table 10.9 (females) to determine percentage (%) body fat from the sum of four skin-fold measurements and your age.

What is your % body fat? _____ %

Table 10.8 Percentage Fat Estimates for Males

Sum of Four Skin Folds (mm)	Age								
	18–22	23–27	28–32	33–37	38–42	43–47	48–52	53–57	58–older
8–12	1.9%	2.5%	3.2%	3.8%	4.4%	5.0%	5.7%	6.3%	6.9%
13–17	3.3%	3.9%	4.5%	5.1%	5.7%	6.4%	7.0%	7.6%	8.2%
18–22	4.5%	5.2%	5.8%	6.4%	7.0%	7.7%	8.3%	8.9%	9.5%
23–27	5.8%	6.4%	7.1%	7.7%	8.3%	8.9%	9.5%	10.2%	10.8%
28–32	7.1%	7.7%	8.3%	8.9%	9.5%	10.2%	10.8%	11.4%	12.0%
33–37	8.3%	8.9%	9.5%	10.1%	10.8%	11.4%	12.0%	12.6%	13.2%
38–42	9.5%	10.1%	10.7%	11.3%	11.9%	12.6%	13.2%	13.8%	14.4%
43–47	10.6%	11.6%	11.9%	12.5%	13.1%	13.7%	14.4%	15.0%	15.6%
48–52	11.8%	12.4%	13.0%	13.6%	14.2%	14.9%	15.5%	16.1%	16.7%
53–57	12.9%	13.5%	14.1%	14.7%	15.4%	16.0%	16.6%	17.2%	17.9%
58–62	14.0%	14.6%	15.2%	15.8%	16.4%	17.1%	17.7%	18.3%	18.9%
63–67	15.0%	15.6%	16.3%	16.9%	17.5%	18.1%	18.8%	19.4%	20.0%
68–72	16.1%	16.7%	17.3%	17.9%	18.5%	19.2%	19.8%	20.4%	21.0%
73–77	17.1%	17.7%	18.3%	18.9%	19.5%	20.2%	20.8%	21.4%	22.0%
78–82	18.0%	18.7%	19.3%	19.9%	20.5%	21.0%	21.8%	22.4%	23.0%
83–87	19.0%	19.6%	20.2%	20.8%	21.5%	22.1%	22.7%	23.3%	24.0%
88–92	19.9%	20.5%	21.2%	21.8%	22.4%	23.0%	23.6%	24.3%	24.9%
93–97	20.8%	21.4%	22.1%	22.7%	23.3%	23.9%	24.8%	25.2%	25.8%
98–102	21.7%	22.6%	22.9%	23.5%	24.2%	24.8%	25.4%	26.0%	26.7%
103–107	22.5%	23.2%	23.8%	24.4%	25.0%	25.6%	26.3%	26.9%	27.5%
108–112	23.4%	24.0%	24.6%	25.2%	25.8%	26.5%	27.1%	27.7%	28.3%
113–117	24.1%	24.8%	25.4%	26.0%	26.6%	27.3%	27.9%	28.5%	29.1%
118–122	24.9%	25.5%	26.2%	26.8%	27.4%	28.0%	28.6%	29.3%	29.9%
123–127	25.7%	26.3%	26.9%	27.5%	28.1%	28.8%	29.4%	30.0%	30.6%
128–132	26.4%	27.0%	27.6%	28.2%	28.8%	29.5%	30.1%	30.7%	31.3%
133–137	27.1%	27.7%	28.3%	28.9%	29.5%	30.2%	30.8%	31.4%	32.0%
138–142	27.7%	28.3%	29.0%	29.6%	30.2%	30.8%	31.4%	32.1%	32.7%
143–147	28.3%	29.0%	29.6%	30.2%	30.8%	31.5%	32.1%	32.7%	33.3%
148–152	29.0%	29.6%	30.2%	30.8%	31.4%	32.7%	32.7%	33.3%	33.9%
153–157	29.5%	30.2%	30.8%	31.4%	32.0%	32.7%	33.3%	33.9%	34.5%
158–162	30.1%	30.7%	31.3%	31.9%	32.6%	33.2%	33.8%	34.4%	35.1%
163–167	30.6%	31.2%	31.9%	32.5%	33.1%	33.7%	34.3%	35.0%	35.6%
168–172	31.1%	31.7%	32.4%	33.0%	33.6%	34.2%	34.8%	35.5%	36.1%
173–177	31.6%	32.2%	32.8%	33.5%	34.1%	34.7%	35.3%	35.9%	36.6%
178–182	32.0%	32.7%	33.3%	33.9%	34.5%	35.2%	35.8%	36.4%	37.0%
183–187	32.5%	33.1%	33.7%	34.3%	34.9%	35.6%	36.2%	36.8%	37.4%
188–192	32.9%	33.5%	34.1%	34.7%	35.3%	36.0%	36.6%	37.2%	37.8%
193–197	33.2%	33.8%	34.5%	35.1%	35.7%	36.3%	37.0%	37.8%	38.2%
198–202	33.6%	34.2%	34.8%	35.4%	36.1%	36.7%	37.3%	37.9%	38.5%
203–207	33.9%	34.5%	35.1%	35.7%	36.4%	37.0%	37.6%	38.2%	38.9%

Table 10.9   Percentage Fat Estimates for Females

Sum of Four Skin Folds (mm)	Age								
	18–22	23–27	28–32	33–37	38–42	43–47	48–52	53–57	58–older
8–12	8.8%	9.0%	9.2%	9.4%	9.5%	9.7%	9.9%	10.1%	10.3%
13–17	10.8%	10.9%	11.1%	11.3%	11.5%	11.7%	11.8%	12.0%	12.2%
18–22	12.6%	12.8%	13.0%	13.2%	13.4%	13.5%	13.7%	13.9%	14.1%
23–27	14.5%	14.6%	14.8%	15.0%	15.2%	15.4%	15.6%	15.7%	15.9%
28–32	16.2%	16.4%	16.6%	16.8%	17.0%	17.1%	17.3%	17.5%	17.7%
33–37	17.9%	18.1%	18.3%	18.5%	18.7%	18.9%	19.0%	19.2%	19.4%
38–42	19.6%	19.8%	20.0%	20.2%	20.3%	20.5%	20.7%	20.9%	21.1%
43–47	21.2%	21.4%	21.6%	21.8%	21.9%	22.1%	22.3%	22.5%	22.7%
48–52	22.8%	22.9%	23.1%	23.3%	23.5%	23.7%	23.8%	24.0%	24.2%
53–57	24.2%	24.4%	24.6%	24.8%	25.0%	25.2%	25.3%	25.5%	25.7%
58–62	25.7%	25.9%	26.0%	26.2%	26.4%	26.6%	26.8%	27.0%	27.1%
63–67	27.1%	27.2%	27.4%	27.6%	27.8%	28.0%	28.2%	28.3%	28.5%
68–72	28.4%	28.6%	28.7%	28.9%	29.1%	29.3%	29.5%	29.7%	29.8%
73–77	29.6%	29.8%	30.0%	30.2%	30.4%	30.6%	30.7%	30.9%	31.1%
78–82	30.9%	31.0%	31.2%	31.4%	31.6%	31.8%	31.9%	32.1%	32.3%
83–87	32.0%	32.2%	32.4%	32.6%	32.7%	32.9%	33.1%	33.3%	33.5%
88–92	33.1%	33.3%	33.5%	33.7%	33.8%	34.0%	34.2%	34.4%	34.6%
93–97	34.1%	34.3%	34.5%	34.7%	34.9%	35.1%	35.2%	35.4%	35.6%
98–102	35.1%	35.3%	35.5%	35.7%	35.9%	36.0%	36.2%	36.4%	36.6%
103–107	36.1%	36.2%	36.4%	36.6%	36.8%	37.0%	37.2%	37.3%	37.5%
108–112	36.9%	37.1%	37.3%	37.5%	37.7%	37.9%	38.0%	38.2%	38.4%
113–117	37.8%	37.9%	38.1%	38.3%	39.2%	39.4%	39.6%	39.8%	40.0%
118–122	38.5%	38.7%	38.9%	39.1%	39.4%	39.6%	39.8%	40.0%	40.5%
123–127	39.2%	39.4%	39.6%	39.8%	40.0%	40.1%	40.3%	40.5%	40.7%
128–132	39.9%	40.1%	40.2%	40.4%	40.6%	40.8%	41.0%	41.2%	41.3%
133–137	40.5%	40.7%	40.8%	41.0%	41.2%	41.4%	41.6%	41.7%	41.9%
138–142	41.0%	41.2%	41.4%	41.6%	41.7%	41.9%	42.1%	42.3%	42.5%
143–147	41.5%	41.7%	41.9%	42.0%	42.2%	42.4%	42.6%	42.8%	43.0%
148–152	41.9%	42.1%	42.3%	42.8%	42.6%	42.8%	43.0%	43.2%	43.4%
153–157	42.3%	42.5%	42.6%	42.8%	43.0%	43.2%	43.4%	43.6%	43.7%
158–162	42.6%	42.8%	43.0%	43.1%	43.3%	43.5%	43.7%	43.9%	44.1%
163–167	42.9%	43.0%	43.2%	43.4%	43.6%	43.8%	44.0%	44.1%	44.3%
168–172	43.1%	43.2%	43.4%	43.6%	43.8%	44.0%	44.2%	44.3%	44.5%
173–177	43.2%	43.4%	43.6%	43.8%	43.9%	44.1%	44.3%	44.5%	44.7%
178–182	43.3%	43.5%	43.7%	43.8%	44.0%	44.2%	44.4%	44.6%	44.8%

9. Multiply your present body weight in pounds (lb) by your present % body fat to determine how much of your body weight is fat. For example, if a male weighs 180 lb and his percentage body fat is 20%, then $180 \times .20 = 36$ lb. His body fat is 36 lb.

How much of your body weight is fat? _____ lb

10. Subtract the weight of fat from your present body weight. This is lean body weight (LBW). For example, if a male weighs 180 lb and his body fat is 36 lb, then $180 - 36 = 144$. His LBW is 144 lb.

What is your **LBW?** _____ lb

11. Calculate ideal weight based on body composition.

   For males, 84% of ideal weight should be LBW. Therefore, ideal weight = LBW/0.84. What is your ideal weight based on body composition? For example, if a male has a LBW of 144 lb, his ideal weight is 144/.84 = 171 lb.

   > Male ideal weight = _____ lb

   (At this weight, your body will contain 16% fat, the amount generally recommended for males.)

   For females, 77% of ideal weight should be LBW. Therefore, ideal weight = LBW/0.77. What is your ideal weight based on body composition?

   > Female ideal weight = _____ lb

   (At this weight, your body will contain 23% fat, the amount generally recommended for females.)

Comparison
How does your ideal weight based on body composition compare to your ideal weight based on BMI?

_____

_____

Conclusions: Ideal Body Weight

- What is your average daily energy requirement? _____ kcal (p. 123)

- What is your average daily energy intake? _____ kcal (p. 116)

- What is your ideal weight? (Take an average of any you have calculated.) _____ lb

   For your information, 1 lb of fat represents 3,500 kcal. Therefore, if you want to lose 1 lb of fat, you must either reduce your intake by 3,500 kcal or increase your activity by 3,500 kcal. Assuming the same amount of physical activity, you could lose 1 lb per week by reducing your kcal intake by 500 kcal per day or gain 1 lb per week by increasing your kcal intake by 500 kcal per day.

- What is your recommendation for maintaining or achieving your ideal weight? _____

Application for Daily Living

Losing Weight

Before turning to more drastic measures, an overweight or obese person should first attempt to lose weight by lowering his or her caloric intake and increasing caloric output through exercise. Fad diets and pills lack lasting value. The effort to control weight is worth it because two serious illnesses are associated with obesity: type 2 diabetes and cardiovascular disease.

When a person has type 2 diabetes, the pancreas produces insulin but the cells do not respond to it. Therefore, glucose builds up in the blood and spills over into the urine. How might diet contribute to the occurrence of type 2 diabetes? Simple sugars in foods, such as candy and ice cream, immediately enter the bloodstream, as do sugars from the digestion of starch within white bread and potatoes. This apparently leads to a high blood fatty acid level, insulin resistance, and type 2 diabetes.

Jared Fogle's pants before he started dieting and exercising.

Cardiovascular disease due to arteries blocked by plaque is another condition seen in obese individuals. Plaque contains saturated fats and cholesterol. Therefore, limiting cholesterol (present in cheese, egg yolks, shrimp, and lobster) in the diet may be helpful. You should also be aware that beef, dairy foods, and coconut oil are rich sources of saturated fat. Further, processed foods made with or fried in partially hydrogenated oils (e.g., vegetable shortening and stick margarine) are sources of trans fats that contribute to plaque formation.

Laboratory Review 10

_____ 1. What is a listing of all foods eaten for a day?

_____ 2. What does *kcal* mean?

_____ 3. Daily energy requirement includes energy for physical activity, SDA, and what else?

_____ 4. Give an example of a basal metabolism activity.

_____ 5. SDA refers to energy needed for what activity?

_____ 6. With age, the BMR decreases. What is the implication for daily energy intake?

_____ 7. Why does a tall, thin person have a higher BMR than a short, stout person?

_____ 8. Daily energy requirement must be in balance with what to not gain weight?

_____ 9. If the average energy intake/day equals the average energy requirement/day, the

person will not _____.

_____ 10. Ideal weight includes lean body weight and what else?

_____ 11. Generally speaking, a male should have no more than what percentage body fat?

_____ 12. Calculation of ideal weight is considered to be the most accurate if it is based

on your _____.

_____ 13. How many kcal are equal to a pound of fat?

_____ 14. In which sex is it natural to have a higher percentage of body fat?

Thought Questions

15. Why do males typically have a higher basal metabolic rate than females?

16. Why is energy needed when the body is at rest?

17. Why is energy needed to process food?

18. Why should someone monitor his or her average daily energy intake when attempting to maintain average body weight?

|BIOLOGY

Find even more resources to learn the chapter concepts including animations, tutorial videos, and interactive practice questions.

McGraw-Hill Access Science Website

An online encyclopedia of science and technology that provides information, including videos, that can enhance the laboratory experience.

www.accessscience.com

11

Homeostasis

Learning Outcomes

Introduction
- Define homeostasis and the internal environment.

11.1 Kidneys
- Understand kidney and nephron structure and blood supply.
- State the three steps in urine formation and how they relate to the parts of a nephron.
- Perform a urinalysis and explain how the results are related to kidney functions.

Prelab Question: What substance will be present in urine if a person has diabetes mellitus?

11.2 Lungs
- Describe the mechanics of breathing and the role of the alveoli in gas exchange.
- Explain how the excretion of CO_2 helps stabilize the pH of the blood.
- Measure respiratory volumes (e.g., tidal volume) and explain their relationship to homeostasis.

Prelab Question: When CO_2 exits blood at the alveoli, how does the pH of blood change?

11.3 Liver
- Describe the anatomy of the liver, including the path of blood from the intestines, through the liver and to the heart.
- Compare the glucose level in the mesenteric artery, the hepatic portal vein, and the hepatic vein before and after eating.
- State two ways the liver contributes to homeostasis.

Prelab Question: What hormone causes the liver to store glucose as glycogen?

Application for Daily Living: **The Liver Is a Vital Organ**

Introduction

Homeostasis refers to the dynamic equilibrium of the body's internal environment. The internal environment of humans consists of blood and tissue fluid. To meet their needs, cells take nutrients (e.g., glucose) and O_2 from the blood and return waste products, including CO_2, to the blood by way of tissue fluid. Tissue fluid in turn exchanges molecules with the blood. Homeostasis also involves adjusting blood pH, ionic concentrations, and blood volume. All internal organs contribute to homeostasis, but this laboratory specifically examines the contributions of the kidneys, lungs, and the liver.

Negative Feedback

Two systems in the body, the endocrine (hormone) system and the nervous system, work together to regulate homeostasis, the relative constancy of the internal environment (blood and tissue fluid). Both systems operate by utilizing negative feedback. For example, when you eat a candy bar and your blood glucose rises, an endocrine gland called the pancreas senses the rise in blood sugar and releases the hormone insulin (Fig. 11.1). Insulin causes the liver to store glucose as glycogen, maintaining blood sugar level at a normal level. The drop in blood sugar level is **negative feedback** that causes the pancreas to stop secreting insulin. Similarly, when blood pressure lowers, the brain sends out impulses to the smooth muscles of arterioles to constrict, raising blood pressure. This rise in blood pressure is negative feedback that causes the brain to stop sending nerve impulses to the arterioles; they relax and the blood pressure lowers.

As you study the operation of the kidneys, lungs, and the liver in this laboratory, realize that negative feedback ultimately is what enables these organs to maintain homeostasis.

> **Planning Ahead** Starting the water bath needed for Section 11.3, page 143 now may save you valuable time.

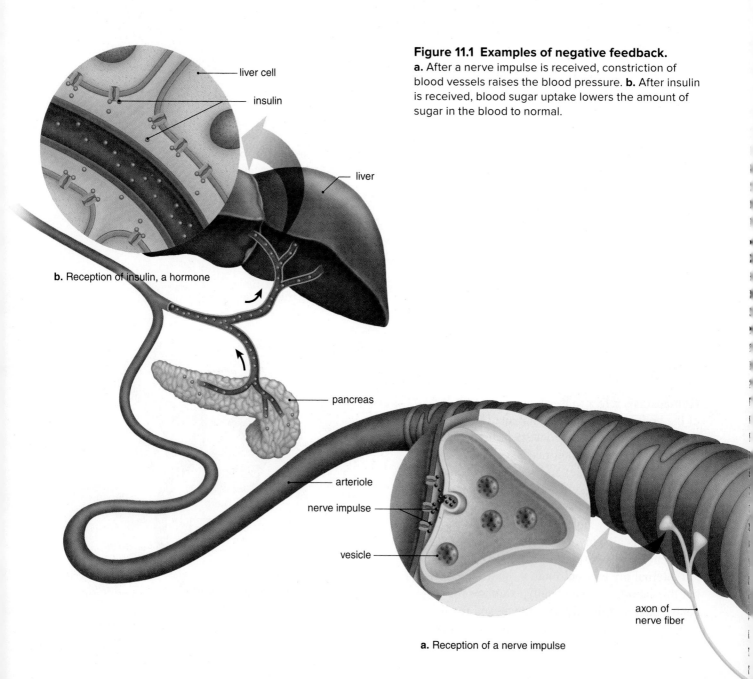

Figure 11.1 Examples of negative feedback.
a. After a nerve impulse is received, constriction of blood vessels raises the blood pressure. **b.** After insulin is received, blood sugar uptake lowers the amount of sugar in the blood to normal.

liver cell

insulin

liver

b. Reception of insulin, a hormone

pancreas

arteriole

nerve impulse

vesicle

axon of nerve fiber

a. Reception of a nerve impulse

Nephron = Basic Unit of Kidney

11.1 Kidneys

In Laboratory 7, we learned that the **kidneys** are bean-shaped organs that lie along the dorsal wall of the abdominal cavity. Now we wish to study kidney structure and function. Figure 11.2 shows the macroscopic and microscopic structure of a kidney. The macroscopic structure of a kidney is due to the placement of over 1 million **nephrons.** Nephrons are tubules that do the work of producing urine.

Figure 11.2   Longitudinal section of a kidney.
a. The kidneys are served by the renal artery and renal vein. **b.** Macroscopically, a kidney has three parts: renal cortex, renal medulla, and renal pelvis. **c.** Microscopically, each kidney contains over a million nephrons.

Calyx

Calyx - Drains Pyramids, into Pelvis

Observation: Kidney Structure

Study a model of a kidney, and with the help of Figure 11.2, locate the following:

1. **Renal cortex:** a granular region that contains most regions of the nephrons
2. **Renal medulla:** contains the renal pyramids consisting of the loops of nephrons and collecting ducts
3. **Renal pelvis:** where urine is received from the collecting ducts

Observation: Nephron Structure

Study the nephron model and with the help of Figure 11.3, identify the following parts of a nephron:

1. **Glomerular capsule** (Bowman's capsule): closed end of the nephron pushed in on itself to form a cuplike structure; the inner layer has pores that allow **glomerular filtration** to occur; substances move from the blood to inside the nephron
2. **Proximal convoluted tubule:** The inner layer of this region has many microvilli that allow tubular reabsorption to occur; substances move from inside the nephron to the blood.

3. **Loop of the nephron:** Nephron narrows to form a U-shaped portion. Functions in water reabsorption.
4. **Distal convoluted tubule:** second convoluted section that lacks microvilli and functions in **tubular secretion;** substances move from blood to inside nephron.

Several nephrons enter one collecting duct. The **collecting ducts** also function in water reabsorption, and they conduct urine to the pelvis of a kidney.

Figure 11.3 Nephron structure and blood supply.
The three main processes in urine formation are described in boxes and color coded to arrows that show the movement of molecules out of or into the nephron at specific locations. In the end, urine is composed of the substances within the collecting duct (see brown arrow).

Study a nephron model and with the help of Figure 11.3 and Table 11.1, trace the path of blood about a nephron:

1. **Afferent arteriole:** small vessel that conducts blood from the renal artery to a nephron.
2. **Glomerulus:** capillary network that exists inside the glomerular capsule; small molecules move from inside the capillary to the inside of the glomerulus during glomerular filtration.
3. **Efferent arteriole:** small vessel that conducts blood from the glomerulus to the peritubular capillary network.
4. **Peritubular capillary network:** surrounds the proximal convoluted tubule, the loop of the nephron, and the distal convoluted tubule.
5. **Venule** takes blood from the peritubular capillary network to the renal vein.

Table 11.1 Circulation of Blood Around a Nephron

Name of Structure	Significance
Afferent arteriole	Brings arteriolar blood toward glomerular capsule
Glomerulus	Capillary tuft enveloped by glomerular capsule
Efferent arteriole	Takes arteriolar blood away from glomerular capsule
Peritubular capillary network	Capillary bed that envelops the rest of the nephron
Venule	Takes venous blood away from the nephron

Kidney Function

The kidneys produce urine and in doing so help maintain homeostasis in several ways. Urine formation requires three steps: glomerular filtration, tubular reabsorption, and tubular secretion (see Fig. 11.3).

Glomerular Filtration

1. Blood entering the glomerulus contains blood cells, proteins, glucose, amino acids, salts, urea, and water. Blood cells and proteins are too large to pass through the glomerular wall and enter the filtrate.
2. Blood pressure causes small molecules of glucose, amino acids, salts, urea, and water to exit the blood and enter the glomerular capsule. The fluid in the glomerular capsule is called the **filtrate**.
3. In the list that follows, *draw an arrow* from left to right for the small substances that leave the glomerulus and become a part of the filtrate.

   Glomerulus   **Glomerular Filtrate**
   Cells
   Proteins
   Glucose
   Amino acids
   Urea
   Water and salts

4. Complete the second and third columns in Table 11.2, p. 137. Use an X to indicate that the substance is at the locations noted.

Tubular Reabsorption

1. When the filtrate enters the proximal convoluted tubule, it contains glucose, amino acids, urea, water, and salts. Some water and salts remain in the nephron but enough are passively reabsorbed into the peritubular capillary to maintain blood volume and blood pressure. Use this information to state a way kidneys help maintain homeostasis. _____

_____

2. The cells that line the proximal convoluted tubule are also engaged in active transport and usually completely reabsorb nutrients (glucose and amino acids) into the peritubular capillary. What would happen to cells if the body lost all its nutrients by way of the kidneys? _____

_____

3. Which of the filtrate substances is reabsorbed the least and will become a part of urine? _____

Urea is a nitrogenous waste. State here another way kidneys contribute to homeostasis. _____

_____

4. In the list that follows, *draw an arrow* from left to right for all those molecules passively reabsorbed into the blood of the peritubular capillary. Use darker arrows for those that are reabsorbed completely by active transport.

Proximal Convoluted Tubule   **Peritubular Capillary**

Water and salts

Glucose

Amino acids

Urea

Tubular Secretion

During tubular secretion, certain substances—for example, penicillin and histamine—are actively secreted from the peritubular capillary into the fluid of the tubule. Also, hydrogen ions (H^+) and ammonia (NH_3) are secreted as NH_4^+ as necessary.

The excretion of H^+ in this way raises the pH of the blood. State here a third way the kidneys contribute to homeostasis. _____

We learned earlier that the lungs help raise the pH of the blood by excreting CO_2 but only the kidneys can excrete H^+.

Summary: Kidney Function and Homeostasis

For each substance listed at the left in Table 11.2, place an X in the last column if you expect the substance to be present in urine.

Answer the following questions.

1. The presence of urea in the urine illustrates which of the kidney's functions? _____

Do the kidneys make urea? _____ What organ makes urea? _____

What do the kidneys produce? _____

2. The presence of NH_4^+ in the urine illustrates which of the kidney's functions? _____

_____

3. Regulation of the blood's water and salt content by the kidneys helps maintain _____

_____ within normal limits.

Table 11.2   Urine Constituents

Substance	In Blood of Glomerulus	In Filtrate	In Urine
Protein (albumin)			
Glucose and amino acids			
Urea			
Water and salts			
NH_4^+			

Urinalysis: A Diagnostic Tool

Urinalysis can indicate whether the kidneys are functioning properly or whether an illness such as diabetes mellitus is present. The procedure is easily performed with a Chemstrip test strip, which has indicator spots that produce specific color reactions when certain substances are present in the urine.

Experimental Procedure: Urinalysis

A urinalysis has been ordered, and you are to test the urine. (In this case, you will be testing simulated urine, just as you tested simulated blood sera earlier in this lab.)

Assemble Supplies

1. Obtain three Chemstrip urine test strips, each of which tests for leukocytes, pH, protein, glucose, ketones, and blood, as noted in Figure 11.4.
2. The color key on the diagnostic color chart or on the Chemstrip vial label will explain what any color changes mean in terms of the pH level and amount of each substance present in the urine sample. You will use these color blocks to read the results of your test.
3. Obtain three "specimen containers of urine" marked 1 through 3. Among them is a normal specimen and two that indicate the patient has an illness.

Tests For:	Normal	1	2	3
leukocytes	negative	Trace	trace	trace
pH	pH 5	5.0	6	6
protein	negative	Trace	Trace	100
glucose	normal	–	2000	–
ketones	negative	–	pos	–
blood	negative	mod/lg	mod/lg	sm

(handwritten: Filtrate is not urine! └ made first   └ end product @ end of process)

Figure 11.4   Urinalysis test.
A Chemstrip test strip can help determine illness in a patient by detecting substances in the urine. The normal test results are given for comparative purposes.

Test the Specimen

1. Be sure the chemically treated patches on the test strip are totally immersed. Briefly (no longer than 1 second) dip a test strip into the first specimen of urine.
2. Draw the edge of the strip along the rim of the specimen container to remove excess urine.
3. Turn the test strip on its side, and tap once on a piece of absorbent paper to remove any remaining urine and to prevent the possible mixing of chemicals.
4. After 60 seconds, read the results as follows: Hold the strip close to the color blocks on the diagnostic color chart (Fig. 11.4) or vial label, and match carefully, ensuring that the strip is properly oriented to the color chart. Enter the test results in Figure 11.4. Use a negative symbol (−) for items that are not present in the urine, a plus symbol (+) for those that are present, and a number for the pH.
5. Test the other two specimens. Add these results to Figure 11.4.

Conclusion: Urinalysis

- State below if the urinalysis is normal or indicates a urinary tract infection (leukocytes, blood, and possibly protein in the urine) or the patient has diabetes mellitus (glucose in the urine).

  Test results 1 _____

  Test results 2 _____

  Test results 3 _____

- The hormone insulin promotes the uptake of glucose by cells. When glucose is in the urine, either the pancreas is not producing insulin (diabetes mellitus type 1) or cells are resistant to insulin (diabetes mellitus type 2). Ketones (acids) are also in the urine because the cells are metabolizing fat instead of glucose.

  Explain why cells are metabolizing fat. _____

  _____

  Why is the pH of urine lower than normal? _____

  _____

- If urinalysis shows that proteins are excreted instead of retained in the blood, would capillary exchange in the tissues be normal? _____ Why or why not? _____

  _____

11.2 Lungs

In Laboratory 7, we learned that air moves from the nasal passages to the trachea, bronchi, bronchioles, and finally, lungs. The right and left lungs lie in the thoracic cavity on either side of the heart. A **lung** is a spongy organ consisting of irregularly shaped air spaces called **alveoli** (sing., *alveolus*). The alveoli are lined with a single layer of squamous epithelium and are supported by a mesh of fine, elastic fibers. The alveoli are surrounded by a rich network of tiny blood vessels called **pulmonary capillaries.**

Observation: Lung Structure

1. Observe a prepared slide of a stained section of a lung. In stained slides, the nuclei of the cells forming the thin alveolar walls appear purple or dark blue (Fig. 11.5a).
2. Look for areas with groups of red- or orange-colored, disk-shaped **erythrocytes** (red blood cells). When these appear in strings, you are looking at capillary vessels in side view.
3. In some part of the slide, you may even observe an artery. Thicker, circular or oval structures with a lumen (cavity) are cross sections of **bronchioles,** tubular pathways through which air reaches the air spaces.

simple squamous epithelial cells

alveolus

250×

a. Lung tissue

blood flow

blood flow

alveolus

alveolar wall

pulmonary capillary

b. Alveolus

Figure 11.5   Gas exchange in the lungs.

a. A photomicrograph shows that the lungs contain many air sacs called alveoli. The alveoli are surrounded by pulmonary capillaries. **b.** During gas exchange, carbon dioxide leaves the blood and enters the alveoli; oxygen leaves the alveoli and enters the blood.

Lung Function

1. Lungs carry out gas exchange.

During gas exchange in the lungs, carbon dioxide (CO_2) leaves the blood within the pulmonary capillaries, and oxygen (O_2) leaves the alveoli and enters the blood. *Show gas exchange in Figure 11.5b by writing O_2 or CO_2 on the lines provided.* After the lungs take up O_2, it is transported by hemoglobin (Hb) inside red blood cells as HbO_2. In the tissues, Hb releases O_2 (Fig. 11.6). Why do cells require oxygen?_____

_____

2. The lungs help maintain the pH of the blood.

Carbon dioxide is carried in the blood as bicarbonate ions:

$$CO_2 \quad + \quad H_2O \longrightarrow H_2CO_3 \longrightarrow HCO_3^- \quad + \quad H^+$$

carbonic acid   bicarbonate ion   hydrogen ions

Hydrogen ions increase the acidity of blood. Is blood more acidic when it is carrying carbon

dioxide? _____ Explain your answer. _____

_____

As CO_2 leaves the blood, this reaction reverses:

$$HCO_3^- \quad + \quad H^+ \longrightarrow H_2CO_3 \longrightarrow CO_2 \quad + \quad H_2O$$

Is blood less acidic when the carbon dioxide exits? _____ Explain your answer. _____

_____

During breathing cessation, it is the increase in acidity rather than the depletion of oxygen that first stimulates the urge to breathe. Chemoreceptors located in the aorta and carotid arteries detect changes in the blood's acidity and stimulate the respiratory center, which triggers contractions in the diaphragm and thoracic muscles, causing alterations in the rate and depth of breathing.

Carbon dioxide transports in Blood as bicarbonate

Figure 11.6 Exchange of gases at the pulmonary capillaries and at the systemic capillaries.

Summary: Lung Function and Homeostasis

1. Gas exchange assists homeostasis because it supplies cells with _____ needed for _____ and rids the body of _____, a metabolic waste.

2. When the lungs excrete CO_2, they help maintain the _____ of the blood. This is the second way that the lungs help maintain homeostasis.

Human Respiratory Volumes

Breathing in, called **inspiration** or inhalation, is the active part of breathing because that's when contraction of rib cage muscles causes the rib cage to move up and out, and contraction of the diaphragm causes the diaphragm to lower. Due to an enlarged thoracic cavity, air is drawn into the lungs. Breathing out, called **expiration** or exhalation, occurs when relaxation of these same muscles causes the thoracic cavity to resume its original capacity. Now air is pushed out of the lungs (Fig. 11.7).

Figure 11.7 Inspiration and expiration.

a. Inspiration occurs after the rib cage moves up and out and the diaphragm moves down. Air rushes in because of the expanded thoracic cavity. **b.** Expiration occurs as the rib cage moves down and in and the diaphragm moves up. As the thoracic cavity gets smaller, air is pushed out.

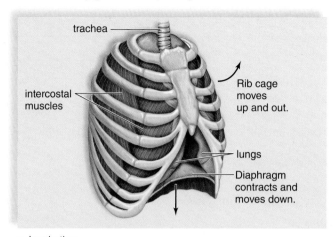

a. Inspiration

b. Expiration

Experimental Procedure: Human Respiratory Volumes

During this Experimental Procedure you will be working with a spirometer, an instrument that measures the amount of exhaled air (Fig. 11.8). Normally, about 500–600 ml of air move into and out of the lungs with each breath. This is called the **tidal volume** (**TV**). You can inhale deeply after a normal breath and more air will enter the lungs; this is the **inspiratory reserve volume (IRV)**. You can also force more air out of your lungs after a normal breath; this is the **expiratory reserve volume (ERV)**. **Vital capacity** is the volume of air that can be forcibly exhaled after forcibly inhaling. →would collapse lungs

Figure 11.8 Nine-liter student wet spirometer.

Tidal Volume (TV)

1. When it's your turn to use the spirometer, install a new disposable mouthpiece and set the spirometer to zero.
2. Inhale normally, then exhale normally (with no extra effort) through the mouthpiece of the spirometer. Record your measurement in Table 11.3.

Table 11.3 Measurements of Lung Volumes

Tidal Volume (TV) sitting	Expiratory Reserve Volume (ERV) Stand	Vital Capacity (VC)	Inspiratory Reserve Volume (IRV)
1st 1000	1st 2400	1st 3900	—— 500
2nd 1500	2nd 3100	2nd 4300	—— ⁻300
3rd 1700	3rd 3500	3rd 4000	—— ~1200
Average 1400 ml	Average 3000 ml	Average 4066 ml	Calculated value = ⁻334 ml

$$IRV = VC - (TV + ERV) = \underline{\qquad} ml$$

3. Three readings are needed, so twice more set the spirometer to zero and repeat the same procedure. Record your measurements in Table 11.3.
4. Later, if necessary, change your readings to milliliters (ml), and calculate your average TV in ml.

In your own words, what is tidal volume?

_____

_____

Expiratory Reserve Volume (ERV)

1. Make sure the spirometer is set to zero.
2. Inhale and exhale normally and then force as much air out as possible into the spirometer. Record your measurement in Table 11.3.
3. Three readings are needed, so twice more set the spirometer to zero and repeat the same procedure. Record your measurements in Table 11.3.
4. Later, if necessary, change your readings to ml, and calculate your average ERV.

In your own words, what is expiratory reserve volume? _____

_____

Vital Capacity (VC)

1. Make sure the spirometer is set to zero.
2. Inhale as much as possible and then exhale as much as possible into the spirometer.
3. Three readings are needed, so twice more set the spirometer to zero and repeat the same procedure. Record your measurements in Table 11.3.
4. Later, if necessary, change your readings to ml, and calculate your average VC.

In your own words, what is vital capacity? _____

_____

Inspiratory Reserve Volume (IRV)

It will be necessary for us to calculate IRV because a spirometer only measures exhaled air, not inhaled air.

Explain. _____

From having measured vital capacity (VC) you can see that VC = TV + IRV + ERV. To calculate IRV, simply subtract the average TV + the average ERV from the value you recorded for the average VC:

   IRV = VC − (TV + ERV) = _____ ml. Record your IRV in Table 11.3.

Conclusions: Human Respiratory Volumes

• Vital capacity varies with age, sex, and height; however, typically for men vital capacity is about 5,200 ml and for women it is about 4,000 ml. How does your vital capacity compare to the typical values for your

gender? _____ If smaller than normal, are you a smoker or is there any health reason why it would be smaller? If larger than normal, are you a sports enthusiast or do you play

a musical instrument that involves inhaling and exhaling deeply? _____
• Diffusion alone accounts for pulmonary gas exchange. Therefore, how does good lung ventilation assist gas

exchange? _____

_____

11.3 Liver

(handwritten: Blood pools in liver, Liver processes what's in the blood.)

As we learned in Laboratory 7, the **liver,** which is the largest organ in the body, lies mainly to the right under the diaphragm. The liver has two main **lobes** and the lobes are further divided into **lobules,** which contain the cells of the liver, called **hepatic cells** (Fig. 11.9a).

Observation: Liver Structure

Study a model of the liver, and identify the following:

1. **Right and left lobes:** The liver has two lobes. The right lobe is larger than the left lobe. Each lobe has many **lobules.** Each lobule has many cells.
2. **Hepatic artery:** the blood vessel that transports O_2-rich blood to the liver
3. **Hepatic portal vein:** the blood vessel that transports blood containing nutrients from the intestine to the liver (Fig. 11.9b).
4. **Hepatic veins:** the blood vessels that transport O_2-poor blood out of the liver to the inferior vena cava

Liver Function

The liver has many functions in homeostasis, and this lab will study two of these.

1. The liver produces urea.

The liver removes amino groups ($—NH_2$) from amino acids and converts them to urea. **Urea** is the primary nitrogenous end product of metabolism that is excreted by the kidneys. What remains of the amino acid (a hydrocarbon) can be used by the body in a number of ways.

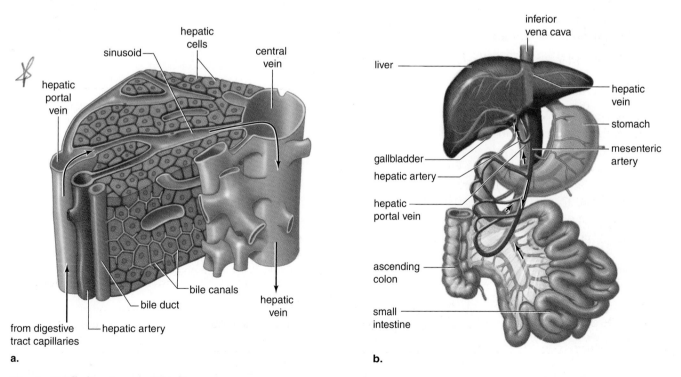

a.

b.

Figure 11.9 Anatomy of the liver.
a. Liver lobule. **b.** The hepatic portal vein lies between the digestive tract and the liver. The hepatic portal vein lies between the digestive tract and the liver. The hepatic veins enter the inferior vena cava which returns blood to the heart.

In the chemical formula for urea that follows, circle the portions that would have come from amino groups:

$$O$$
$$\|$$
$$NH_2-C-NH_2$$

2. The liver regulates the blood glucose level.

After you eat, blood glucose level rises. This increase is detected by the pancreas, which in response releases the hormone insulin. Insulin causes the liver to store excess glucose as glycogen. Before your next meal, the drop in blood sugar causes the pancreas to release glucagon, a hormone that promotes the liver to break down glycogen and release glucose.

Complete the following equation by writing *glucose* and *glycogen* on the appropriate sides of the arrows.

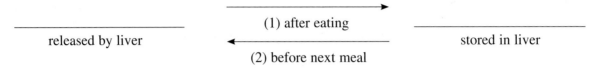

released by liver   (1) after eating   stored in liver

(2) before next meal

If glucose is excreted in the urine, instead of being stored, the individual has the medical condition called **diabetes mellitus,** commonly known as diabetes. In type 1 diabetes, the pancreas is no longer making insulin; in type 2 diabetes, the plasma membrane receptors are unable to bind properly to insulin. In type 1 diabetes, but not type 2, ketones (strong organic acids), a breakdown product of fat metabolism, also appear in the urine.

Experimental Procedure: Blood Glucose Level After Eating

Study the diagram of the human cardiovascular system in Figure 8.7 (p. 101), and trace the path of blood from the mesenteric artery to the vena cava via the intestine and liver. Simulated serum samples have been prepared to correspond to these blood vessels in a person who ate a short time ago:

A_1: Serum from a mesenteric artery. The mesenteric arteries take blood from the aorta to the intestine.

B_1: Serum from the hepatic portal vein, which lies between the intestine and the liver.

C_1: Serum from the hepatic vein, which takes blood from the liver to the vena cava.

> ⚠ **Benedict's reagent** Use protective eyeware when performing this experiment. Benedict's reagent is highly corrosive. Exercise care in using this chemical. If any should spill on your skin, wash the area with mild soap and water. Follow your instructor's directions for disposal of this chemical.

1. With a wax pencil, label three test tubes A_1, B_1, and C_1, and mark them at 1 cm and 2 cm.
2. Fill Tube A_1 to the 1 cm mark with serum A_1 and to the 2 cm mark with Benedict's reagent.
3. Fill Tube B_1 to the 1 cm mark with serum B_1 and to the 2 cm mark with Benedict's reagent.
4. Fill Tube C_1 to the 1 cm mark with serum C_1 and to the 2 cm mark with Benedict's reagent.
5. Place all three test tubes in the water bath at the same time. Heat the tubes in the same boiling water bath for 5 minutes.
6. Note any color change in the test tubes, and record the color and your conclusions in Table 11.4. The tube that shows color first has the most glucose, and so forth. Use the following chart to assist you in making your conclusions:

Color Change	Amount of Glucose
Color is still blue.	None
Green	Very low
Yellow-orange	Moderate
Orange	High
Orange-red	Very high

Table 11.4 Blood Glucose Level After Eating

Test Tubes	Color (After Heating)	Conclusion
A$_1$ (mesenteric artery)	orange	high
B$_1$ (hepatic portal vein)	orange red	very high
C$_1$ (hepatic vein)	yellow orange	moderate

Conclusions: Blood Glucose Level After Eating

- Which blood vessel—a mesenteric artery, the hepatic portal vein, or the hepatic vein—contains the most glucose after eating? _B1 hepatic portal vein_

- Why do you suppose that the hepatic vein does not contain as much glucose as the hepatic portal vein after eating? _____

Experimental Procedure: Blood Glucose Level Before Next Meal

Simulated serum samples have been prepared to correspond to these blood vessels in a person who has not eaten for some time:

A$_2$: Serum from a mesenteric artery
B$_2$: Serum from the hepatic portal vein
C$_2$: Serum from the hepatic vein

1. With a wax pencil, label three test tubes A$_2$, B$_2$, and C$_2$, and mark them at 1 cm and 2 cm.
2. Fill Tube A$_2$ to the 1 cm mark with serum A$_2$ and to the 2 cm mark with Benedict's reagent.
3. Fill Tube B$_2$ to the 1 cm mark with serum B$_2$ and to the 2 cm mark with Benedict's reagent.
4. Fill Tube C$_2$ to the 1 cm mark with serum C$_2$ and to the 2 cm mark with Benedict's reagent.
5. Heat the tubes in the same boiling water bath for 5 minutes.
6. Note any color change in the test tubes, and record the color and your conclusions in Table 11.5. Use the following list to assist you in making your conclusions:

Color Change	Amount of Glucose
Color is still blue.	None
Green	Very low
Yellow-orange	Moderate
Orange	High
Orange-red	Very high

Table 11.5 Blood Glucose Level Before Next Meal

Test Tubes	Color (After Heating)	Conclusion
A$_2$ (mesenteric artery)	orange	high
B$_2$ (hepatic portal vein)	Blue	no glucose
C$_2$ (hepatic vein)	orange-red	very high

Conclusions: Blood Glucose Level Before Next Meal

- Which blood vessel—a mesenteric artery, the hepatic portal vein, or the hepatic vein—contains the most glucose before your next meal? _____

- Why do you suppose that the hepatic vein now contains more glucose than the hepatic portal vein?

  _____

Summary: Liver Function and Homeostasis

We studied two ways the liver contributes to homeostasis.

1. The liver produces __Urea__, which is excreted by the __Kidney__. Urea is the way we excrete nitrogen from the body; it is a metabolic __waste__.

2. The liver stores glucose as __glycogen__ after eating and releases glucose in between eating so that the concentration of glucose in the blood stays relatively constant at 0.1%. What hormone promotes storage of glucose by the liver? __Insulin__ What hormone promotes breakdown of glycogen to glucose by the liver? __glucogon__

Conclusion: Homeostasis

1. As noted at the beginning of this laboratory, homeostasis is the dynamic equilibrium of the body's internal environment, the blood and tissue fluid surrounding tissue cells. The lungs and kidneys have boundaries that interact with the external environment to refresh blood. The liver also regulates blood content. Fill in the following chart to show the activities of the kidneys, lungs, and liver.

Processes	Kidneys	Lungs	Liver
Gas exchange	_____	O_2 enters and CO_2 exits the blood	_____
pH maintenance	a.	e.	_____
Glucose level	b.	_____	g.
Waste removal	c.	f.	h.
Blood volume	d.	_____	_____

2. Which of these organs contributes most to homeostasis? _____

3. As described throughout this lab, homeostasis is maintained when a change in the internal environment values triggers a response that restores the normal condition. Complete Table 11.6 to show how the kidneys, lungs, and liver specifically respond to changes in the internal environment. Under "Response," also include any hormones involved.

Table 11.6 Changes in Internal Environment

Change	Organ (kidneys, lungs, or liver)	Response
Decrease in blood glucose level		
Decrease in blood volume and pressure		
Increase in blood CO_2		

Application for Daily Living

The Liver Is a Vital Organ

The liver is a vital organ because we can't live without one. The liver is the gatekeeper to the blood. After the molecules enter the blood at the digestive tract, they go first to the liver, which performs these functions.

- As we studied in this lab, the liver maintains glucose in the blood so that cells always have a source of energy. After we eat, the liver stores glucose as glycogen and then it breaks down glycogen gradually so that the blood concentration of glucose stays within a normal range.
- The liver removes and breaks down poisons, i.e., chemicals that can harm the body. The liver destroys old enzymes and hormones that no longer function properly, drugs such as an antibiotics and pain killers that have done their job, and drugs of abuse that damage the brain.
- The liver produces several kinds of blood proteins whose functions are varied. Among these are proteins needed to make the blood clot and if these proteins are missing, a simple cut can cause severe bleeding.
- The liver produces bile, which is sent to the digestive tract, where it emulsifies fat. Without bile, eating ice cream, the icing on a cake, or a juicy steak can make you ill.

It is important to realize that your behavior can affect your liver. The liver ordinarily does a marvelous job but it can be overwhelmed. Alcohol is a poison the liver valiantly breaks down, but if you drink heavily for a long time, the liver can't keep up and develops cirrhosis, characterized by scar tissue that has taken place of working liver cells. Another cause of liver failure is hepatitis, an infection that can be sexually transmitted or acquired by using dirty needles to shoot drugs into the body. When a liver is severely damaged, only a liver transplant can prevent death. The need for a healthy liver provides a good reason for a healthy lifestyle.

P.C.T.   **1.** When molecules leave the glomerulus, they enter what portion of the nephron?

Glucose   **2.** Name a substance that is in the glomerular filtrate but not in the urine.

T. Reabsorb   **3.** Name the process by which molecules move from the proximal convoluted tubule into the blood.

_____ **4.** H^+ is excreted in combination with what molecule?

alveoli   **5.** What are the air spaces in the lungs called?

CO_2 **6.** What molecule is removed by the lungs?

lower/orig capacity **7.** When we exhale, the diaphragm relaxes and moves in what direction?

Normally   **8.** When measuring tidal volume, should a student exhale normally or maximally?

higher   **9.** Vital capacity is expected to have a higher or lower volume than tidal volume?

hepatic portal vein **10.** What blood vessel lies between the intestines and the liver?

glycogen   **11.** In what form is glucose stored in the liver?

urea   **12.** The liver removes the amino group from amino acids to form what molecules?

Thought Questions

13. Smoking cigarettes causes emphysema, in which the alveoli burst. Why would you expect the patient to have an abnormally low vital capacity and be tired? Don't have same capacity of lungs to take in as much air. Would cause person to be out of breath and winded faster than a non smoker

14. Which systemic blood vessel would you expect to have a high glucose content immediately after eating? _____ Explain your answer. hepatic portal vein? First vessel out of the GI tract

15. In what ways do the kidneys aid homeostasis?

12

Musculoskeletal System

Introduction

The human skeletal system consists of the bones (206 in adults) and joints, along with the cartilage and ligaments that occur at the joints. The muscular system contains three types of muscles: smooth, cardiac, and skeletal. The term **musculoskeletal system** recognizes that contraction of skeletal muscles causes the bones to move.

In humans, the skeletal muscles are most often attached across a joint (Fig. 12.1). The biceps brachii muscle has two origins, and the triceps brachii has three origins on the humerus and scapula. Find the tendon of insertion of the biceps brachii muscle by feeling on the anterior surface of your elbow while contracting your biceps muscle. Feel for the bone at your posterior elbow. This is the ulna, the site of insertion of the tendon for the triceps brachii muscle.

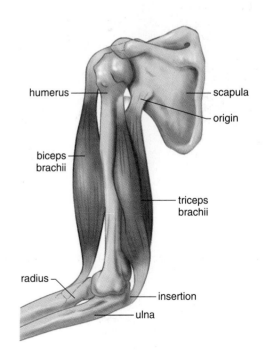

Figure 12.1 Muscular action.
Muscles, such as these muscles of the arm (which have their origin on the scapula and their insertion on the bones of the forearm), cause bones to move.

Muscles work in antagonistic pairs. For example, when the biceps brachii contracts, the bones of the forearm are pulled upward, while the triceps brachii relaxes; when the triceps brachii contracts, the bones of the forearm are pulled downward, while the biceps brachii relaxes.

12.1 Anatomy of a Long Bone

Although the bones of the skeletal system vary considerably in shape as well as in size, a long bone, such as the human femur, illustrates the general principles of bone anatomy (Fig. 12.2).

Figure 12.2   Anatomy of a bone from the macroscopic to the microscopic level.
A long bone is encased by the periosteum except at the ends, where it is covered by hyaline (articular) cartilage (see micrograph, *top left*). Spongy bone located at the ends may contain red bone marrow. The medullary cavity contains yellow bone marrow and is bordered by compact bone, shown in the enlargement and micrograph (*right*).

Hyaline cartilage

matrix

chondrocytes in lacunae

250×

Compact bone

osteocytes in lacunae

concentric lamellae

central canal

100×

growth plate

hyaline cartilage
spongy bone (contains red bone marrow)

compact bone

periosteum

Diaphesis

osteon

blood vessel

medullary cavity (contains yellow bone marrow)

Shaft

Spongy bone

trabeculae

osteoblasts

canaliculus

lacuna

4,700×

osteocyte   nucleus

blood vessels

Epiphesis

Observation: Anatomy of a Long Bone

Examine the exterior and a longitudinal section of a long bone or a model of a long bone, and with the help of Figure 12.2, identify the following:

1. **Periosteum:** tough, fibrous connective tissue covering continuous with the ligament and tendons that anchor bones; the periosteum allows blood vessels to enter the bone and service its cells
2. Expanded portions at each end of the bone (epiphysis) that contain spongy bone
3. Extended portion, or shaft (diaphysis) of a long bone that lies between the epiphyses; walls of diaphysis are compact bone
4. **Hyaline (articular) cartilage:** layer of cartilage where the bone articulates with (meets) another bone; decreases friction between bones during movement
5. **Medullary cavity:** cavity located in the diaphysis that stores yellow marrow, which contains a large amount of fat.
   Label the diaphysis and the epiphysis (twice) in Figure 12.2. Which of these contains the growth line where a long bone can grow in length? _____

Observation: Tissues of a Long Bone

The medullary cavity is bounded at the sides by **compact bone** and at the ends by **spongy bone.** Beyond a thin shell of compact bone is the layer of articular cartilage. **Red marrow,** a specialized tissue that produces all types of blood cells, occurs in the spongy bone of the skull, ribs, sternum, and vertebrae and in the ends of the long bones.

1. Examine a prepared slide of compact bone, and with the help of Figure 12.2, identify
   a. **Osteons:** cylindrical structural units
   b. **Lamellae:** concentric rings of matrix
   c. **Matrix:** nonliving material maintained by osteocytes; contains mineral salts (notably calcium salts) and protein
   d. **Lacunae:** cavities between the lamellae that contain osteocytes (bone cells)
   e. **Central canal:** canal in the center of each osteon; Figure 12.2 shows that there are _____ in a central canal.
   f. **Canaliculi:** tiny channels that contain the processes of cells; these processes allow nutrients to pass between the osteocytes (sing., canaliculus)
   Describe how an osteocyte located near a central canal can pass nutrients to osteocytes located far from the central canal. _____
   _____

2. Examine a prepared slide of spongy bone, and with the help of Figure 12.2, identify
   a. **Trabeculae:** bony bars and plates made of mineral salts and protein
   b. **Lacunae:** cavities scattered throughout the trabeculae that contain osteocytes
   c. **Red bone marrow:** within large spaces separated by the trabeculae
   What activity occurs in red bone marrow? _____
   _____

3. Examine a prepared slide of hyaline cartilage, and with the help of Figure 12.2, identify
   a. **Lacunae:** cavities in twos and threes scattered throughout the matrix, which contain chondrocytes (cells that maintain cartilage)
   b. **Matrix:** material more flexible than bone because it consists primarily of protein
   Seniors tend to have joints that creak. What might be the matter? _____

12.2 The Skeleton

The human skeleton is divided into axial and appendicular components (Fig. 12.3). The **axial skeleton** is the main longitudinal portion and includes the skull, the vertebral column, the sternum, and the ribs. The **appendicular skeleton** includes the bones of the appendages and their supportive pectoral and pelvic (shoulder and hip) girdles.

Figure 12.3   The axial and appendicular skeletons.
a. Axial skeleton bones are colored blue. **b.** Bones of the appendicular skeleton are tan.

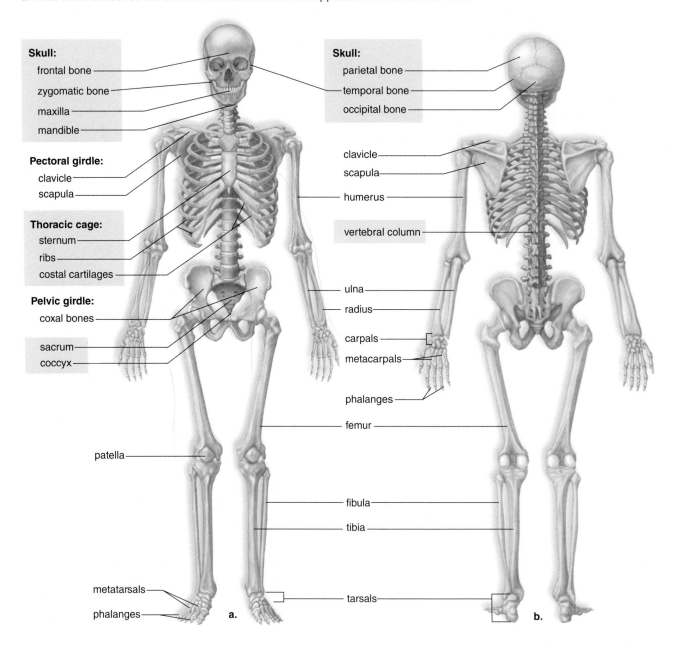

Skull:
   frontal bone
   zygomatic bone
   maxilla
   mandible

Pectoral girdle:
   clavicle
   scapula

Thoracic cage:
   sternum
   ribs
   costal cartilages

Pelvic girdle:
   coxal bones
   sacrum
   coccyx

patella

metatarsals
phalanges

a.

Skull:
   parietal bone
   temporal bone
   occipital bone

clavicle
scapula
humerus

vertebral column

ulna
radius

carpals
metacarpals

phalanges

femur

fibula
tibia

tarsals

b.

Examine a human skeleton, and with the help of Figure 12.4, identify the **foramen magnum,** a large opening through which the spinal cord passes, and the following bones:

1. The **skull** is composed of many bones fused together at fibrous joints called **sutures.** Note the following in the cranium:

    a. **Frontal bone:** forms forehead

    b. **Parietal bones:** extend to sides of skull

    c. **Occipital bone:** curves to form base of skull

    d. **Temporal bones:** located on sides of skull

    e. **Sphenoid bone:** helps form base and sides of skull, as well as part of the orbits

    Which of these bones contribute to forming the face? _____

    Which could best be associated with wearing glasses? _____

Figure 12.4   Skull.
a. Lateral view. **b.** Inferior view.

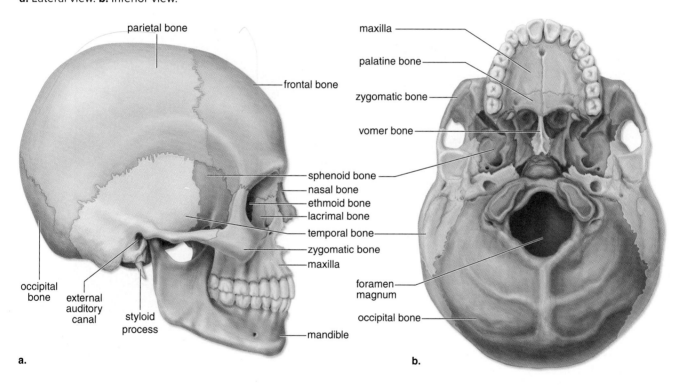

a.

b.

2. Facial bones (Fig. 12.5)

The most prominent of the facial bones are the mandible, the maxillae (sing., maxilla), the zygomatic bones, and the nasal bones. As shown previously, certain of the cranial bones contribute to the face. The frontal bone forms the forehead and thas supraorbital ridges where the eyebrows are located. The temporal bone and the wings of the sphenoid bone account for the flattened areas we call the temples. Note the following:

 a. **Mandible:** the lower jaw

 b. **Maxillae:** the upper jaw and anterior portion of the hard palate

 c. **Palatine bones:** posterior portion of hard palate and floor of nasal cavity

 d. **Zygomatic bones:** cheekbones

 e. **Nasal bones:** bridge of nose

Which of these is movable and allows you to chew your food? _mandible_

Figure 12.5 Bones of the face.

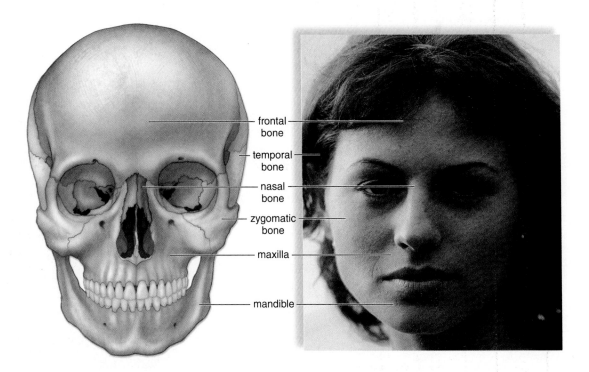

3. The **vertebral column** (Fig. 12.6) provides support and houses the **spinal cord.** It is composed of many vertebrae separated from one another by intervertebral disks. The vertebral column customarily is divided into five series:

a. Seven **cervical vertebrae** (forming the neck region)

b. Twelve **thoracic vertebrae** (with which the ribs articulate)

c. Five **lumbar vertebrae** (in the abdominal region)

d. Five fused sacral vertebrae, called the **sacrum**

e. Four fused caudal vertebrae forming the **coccyx** in humans

Which of the vertebrae can be associated with the chest? _Thoracic_

Which of the vertebrae could be the cause of your aching back? _Lumbar_

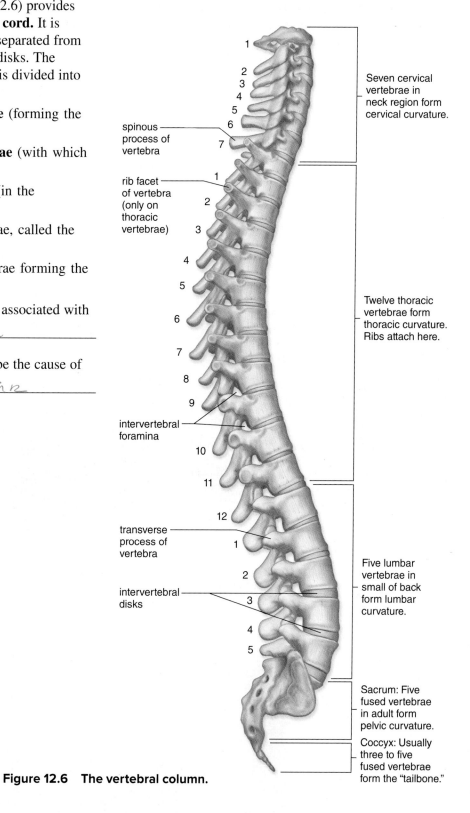

spinous process of vertebra

rib facet of vertebra (only on thoracic vertebrae)

intervertebral foramina

transverse process of vertebra

intervertebral disks

Seven cervical vertebrae in neck region form cervical curvature.

Twelve thoracic vertebrae form thoracic curvature. Ribs attach here.

Five lumbar vertebrae in small of back form lumbar curvature.

Sacrum: Five fused vertebrae in adult form pelvic curvature.

Coccyx: Usually three to five fused vertebrae form the "tailbone."

Figure 12.6   The vertebral column.

4. The twelve pairs of **ribs** and their associated muscles form a bony case that supports the thoracic cavity wall (Fig. 12.7). The ribs connect posteriorly with the thoracic vertebrae, and some are also attached by cartilage directly or indirectly to the sternum. Those ribs without any anterior attachment are called **floating ribs.**

Which of the ribs help form the protective part of the rib cage? _____

Figure 12.7   The ribcage.
a. Art. **b.** Photograph.

a.

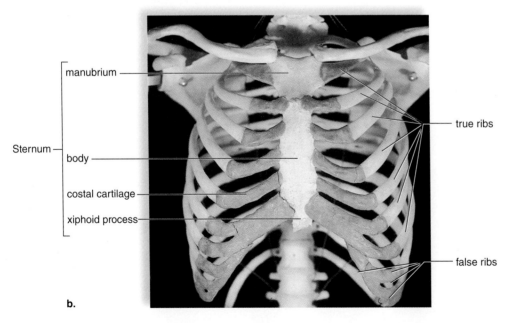

b.

Observation: Appendicular Skeleton

Examine a human skeleton, and with the help of Figure 12.8, identify the following bones:

1. The **pectoral girdles,** which support the upper limbs, are composed of the **clavicle** (collarbone) and **scapula** (shoulder bone).

   Why is the shoulder apt to become dislocated? Ball in socket jt.

2. The upper limb (arm plus the forearm) is composed of the following:

   a. **Humerus:** the large long bone of the arm

   b. **Radius:** the long bone of the forearm, with a pivot joint at the elbow that allows rotational motion

   c. **Ulna:** the other long bone of the forearm, with a hinge joint at the elbow that allows motion in only one plane. Take hold of your elbow, and twist the forearm to show that the radius rotates over the ulna but the ulna doesn't move during this action.

   d. **Carpals:** a group of small bones forming the wrist

   e. **Metacarpals:** slender bones forming the palm

   f. **Phalanges:** the bones of the fingers

   Which of these bones would you use to pick up a teacup? metacarpals / phalanges

clavicle
acromion process
coracoid process
greater tubercle
glenoid cavity
scapula
deltoid tuberosity
humerus
capitulum
head of radius
trochlea
radius
ulna
head of ulna
carpals
metacarpals
phalanges

Figure 12.8   The bones of the pectoral girdle and upper limb.

3. The **pelvic girdle** forms the basal support for the lower limbs and is composed of two **coxal** (hip) **bones** (Fig. 12.9). Each coxal bone consists of the ilium, which is superior to the other two: the pubis and the ischium. (The pubis is ventral to the ischium.)

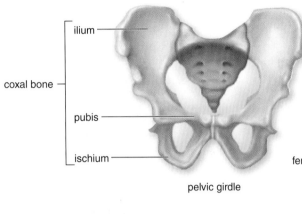

pelvic girdle

The female pelvis is much broader and shallower than that of the male. The angle between the pubic bones looks like a U in females and a V in males. How is it advantageous for the female pelvis to be broader and more shallow than that of a male? _Child birth_

4. The lower limb (the thigh plus the leg) is composed of a series of loosely articulated bones, including the following:

  a. **Femur:** the long bone of the thigh

  b. **Patella:** kneecap

  c. **Tibia:** the larger of the two long bones of the leg; feel for the bump on the inside of the ankle

  d. **Fibula:** the smaller of the two long bones of the leg; feel for the bump on the outside of the ankle

  e. **Tarsals:** a group of small bones forming the ankle

  f. **Metatarsals:** slender anterior bones of the foot

  g. **Phalanges:** the bones of the toes

  Which of these bones would you use to kick a soccer ball? _metatarsals_

Figure 12.9 **The bones of the pelvic girdle and lower limb.**

12.3 The Skeletal Muscles

This laboratory is also concerned with skeletal muscles—those muscles that make up the bulk of the human body. With the help of Figure 12.10, identify the major muscles of the body.

Naming Muscles

Muscles are named for various characteristics, as shown in the following list:

1. **Size:** The gluteus maximus is the largest muscle, and it forms the buttocks.
2. **Shape:** The deltoid is shaped like a Greek letter delta, or triangle.
3. **Direction of fibers:** The rectus abdominis is a longitudinal muscle of the abdomen (*rectus* means "straight").
4. **Location:** The frontalis overlies the frontal bone.
5. **Number of attachments:** The biceps brachii has two attachments, or origins.
6. **Action:** The extensor digitorum extends the fingers, or digits.

Figure 12.10   Human musculature.
Superficial skeletal muscles in (**a**) anterior and (**b**) posterior view.

Orbicularis oculi: blinking, winking, responsible for crow's feet

Orbicularis oris: "kissing" muscle

Pectoralis major: brings arm forward and across chest

Serratus anterior: pulls the scapula (shoulder blade) forward, as in pushing or punching

External oblique: compresses abdomen; rotation of trunk

Quadriceps femoris: straightens leg at knee; raises thigh

Tibialis anterior: turns foot upward, as when walking on heels

Masseter: a chewing muscle; clinches teeth

Deltoid: brings arm away from the side of body; moves arm up and down in front

Biceps brachii: bends forearm at elbow

Rectus abdominis: bends vertebral column; compresses abdomen

Flexor carpi group: bends wrist and hand

Adductor longus: moves thigh toward midline; raises thigh

Sartorius: moves the thigh away from the midline; raises and rotates leg close to body; these combined actions occur when "crossing legs" or soccer kick

Extensor digitorum longus: raises toes; raises foot

Trapezius: raises scapula, as when shrugging shoulders; pulls head backward

Latissimus dorsi: brings arm down and backward behind the body

Triceps brachii: straightens forearm at elbow

Extensor carpi group: straightens wrist and hand

Extensor digitorum: straightens fingers and wrist

Gluteus maximus: extends thigh back

Biceps femoris: bends leg at knee; extends thigh back

Gastrocnemius: turns foot downward, as when standing on toes; bends leg at knee

Achilles tendon

a.

b.

Limbs	
arm: above the elbow	thigh: above the knee
forearm: below the elbow	leg: below the knee

Match these muscles to these functions from Fig. 12.10.

Doing a pelvic tilt and curving the spine _____

Swinging movements of arms during walking and swimming _Rotation_

Pulls forearm toward you when rowing _Flex_____

Helps maintain the trunk in an erect posture _____

Extends and separates the fingers _Abduct_____

Wrinkles the skin of the forehead _____

Joint Movements

Figure 12.11 demonstrates the following types of joint movements:

Flexion   Moving jointed body parts toward each other
Extension   Moving jointed body parts away from each other
Adduction   Moving a part toward a vertical plane running through the longitudinal midline
                     of the body
Abduction   Moving a part away from a vertical plane running through the longitudinal midline
                     of the body
Rotation   Moving a body part around its own axis; **circumduction** is moving a body part
                     in a wide circle.
Inversion   A movement of the foot in which the sole is turned inward
Eversion   A movement of the foot in which the sole is turned outward

Figure 12.11   Joint movements.
a. Flexion and extension. **b.** Adduction and abduction. **c.** Rotation and circumduction. **d.** Inversion and eversion.

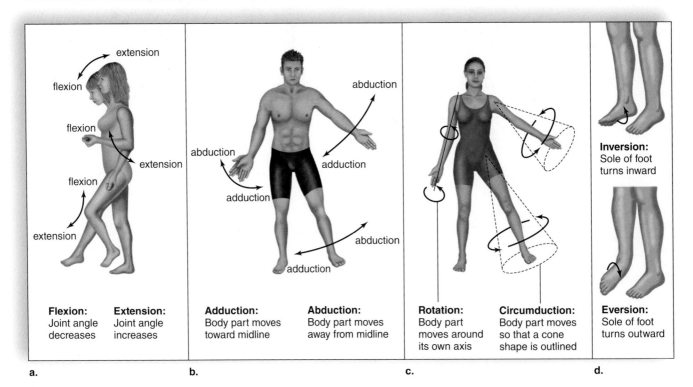

Antagonistic Pairs

Skeletal muscles are attached to the skeleton, and their contraction causes the movement of bones at a joint. Muscles shorten when they contract, so they can only pull; they cannot push. Therefore, muscles work in **antagonistic pairs.** Usually, contraction of one member of the pair causes a bone to move in one direction, and contraction of the other member of the pair causes the same bone to move in an opposite direction.

Are the muscles shown in Figure 12.1 antagonistic?_____ How so? _____

_____

Observation: Antagonistic Pairs

Locate the following antagonistic pairs in Figure 12.10. In each case, state their opposing actions by inserting one of these functions: *flexes, extends, adducts,* or *abducts.*

1. The biceps brachii _flex_____ the forearm.

   The triceps brachii _extend___ the forearm.

2. The sartorius _abducts__ the thigh.

   The adductor longus _adducts__ the thigh.

3. The quadriceps femoris _extend____ the leg.

   The biceps femoris _Flex_____ the leg.

Isometric and Isotonic Contractions

A muscle contains many muscle fibers. When a muscle contracts, usually some fibers undergo isotonic contraction, and others undergo isometric contraction. When the tension of muscle fibers is sufficient to lift a load, many fibers change length as they lift the load. The muscle contraction is said to be **isotonic** (same tension). In contrast, when the tension of muscle fibers is used only to support rather than lift a load, the muscle contraction is said to be **isometric** (same length). The length of many fibers remains the same, but their tension still changes.

Experimental Procedure: Isometric and Isotonic Contractions

Note: The upper limb is composed of the arm plus the forearm.

Isotonic Contraction

1. Start with your left forearm resting on a table. Watch the anterior surface of your left arm while you slowly bend your elbow and bring your left forearm toward the arm. An isotonic contraction of the biceps brachii produces this movement.

2. If muscle contraction produces movement, is this an isometric or isotonic contraction? _____

   _____

Isometric Contraction

1. Place the palm of your left hand underneath a tabletop. Push up against the table while you have your right hand cupped over the anterior surface of your left arm so that you can feel the muscle there undergo an isometric contraction.

2. Is the biceps brachii or the triceps brachii located on the anterior surface of the arm?
biceps brachii

3. What change did you notice in the firmness of this muscle as it contracted? _____
_____

4. Did your forearm move as you pushed up against the table? _____

5. Given your answer to question 4, did this muscle's fibers shorten as you pushed up against the
tabletop? _____

12.4 Mechanism of Skeletal Muscle Fiber Contraction

A whole skeletal muscle is made up of many cells, usually called muscle fibers. Muscle fibers can be examined using a light microscope and using an electron microscope.

Skeletal Muscle Fibers

Muscle fibers are striated—that is, they have alternating light and dark bands. These striations can be observed in a light micrograph of muscle fibers and in an electron micrograph in longitudinal section.

Observation: Skeletal Muscle Tissue

1. Examine a prepared slide of skeletal muscle using a light microscope. Identify the long, multinucleated fibers arranged in a parallel fashion. How do you know the muscle fibers are striated? _____
_____

Skeletal muscle
• has striated cells with multiple nuclei.
• occurs in muscles attached to skeleton.
• functions in voluntary movement of body.

skeletal muscle fiber

striation   nucleus   250×

2. When a skeletal muscle fiber is examined using an electron microscope, it is possible to see that the fibers contain myofibrils which in turn contain actin and myosin filaments (Fig. 12.12). It is the placement of these filaments that cause muscle fibers to be striated.

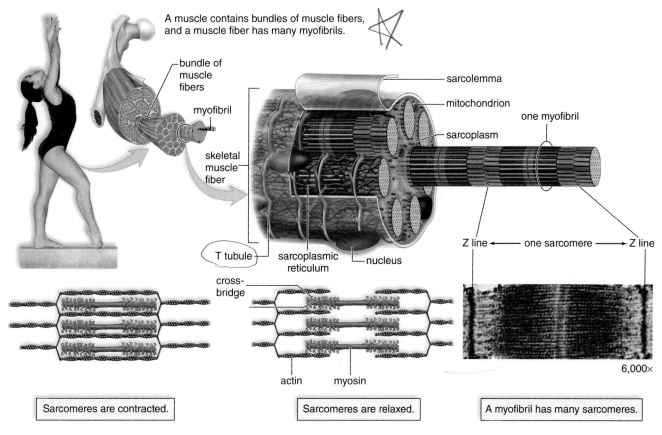

A muscle contains bundles of muscle fibers, and a muscle fiber has many myofibrils.

bundle of muscle fibers

myofibril

skeletal muscle fiber

T tubule

sarcoplasmic reticulum

sarcolemma

mitochondrion

sarcoplasm

one myofibril

nucleus

Z line ← one sarcomere → Z line

6,000×

cross-bridge

actin   myosin

Sarcomeres are contracted.

Sarcomeres are relaxed.

A myofibril has many sarcomeres.

Figure 12.12   Electron microscopy of a skeletal muscle fiber.

Skeletal Muscle Fiber Contraction

Electron microscopy helped investigators determine what causes the striated appearance of a muscle fiber and also its contraction. A muscle fiber contains hundreds, even thousands, of contractile portions called **myofibrils** divided into units called **sarcomeres.** Each sarcomere contains **myosin filaments** and **actin filaments** (Fig. 12.12). The sarcomeres contract when myosin cross-bridges attach to and pull the actin filaments to the center of the sarcomeres.

We are going to study muscle contraction with this data in mind:

1. Myosin is an enzyme that can break down ATP as an energy source for muscle contraction.
2. Enzymes often require cofactors, which can be coenzymes or certain ions, such as the K^+ and Mg^{2+} required by myosin.
3. When ATP, K^+, and Mg^{2+} are present, myosin cross-bridges attach to actin filaments and pull them to the center of sarcomeres within myofibrils. In this way, the actin filaments slide past myosin filaments and the sarcomeres, and a muscle fiber contracts.

During this procedure you are going to expose muscle fibers to different solutions:

1. Glycerol (A viscous liquid that keeps the muscle tissue moist.)
2. K^+ and Mg^{2+} salt solution alone
3. ATP solution alone
4. ATP and salt solution

Hypothesize which of these (1–4) will produce contraction, and explain. _____

_____

The Procedure

1. Label two slides, slide 1 and slide 2. Mount a strand of muscle fibers in a drop of glycerol on each slide. Place each slide on a millimeter ruler, and measure the length of the strand. Record these lengths in the first row in Table 12.1. If there is more than a small drop of glycerol on the slides, soak up the excess on a piece of lens paper held at the edge of the glycerol farthest from the fiber strand.
2. To slide 1, add a few drops of a salt solution containing potassium and magnesium ions (K^+ and Mg^{2+}), and note any change in strand length. Record your results in Table 12.1.
3. To slide 2, add a few drops of ATP solution, and note any change in strand length. Record your results in Table 12.1.
4. Now add ATP solution to slide 1. Note any change in strand length, and record your results in Table 12.1. To slide 2, add a few drops of the K^+/Mg^{2+} salt solution, and note any change in strand length. Record your results in Table 12.1.

Table 12.1   Glycerinated Muscle Contraction

Solution	Length of Muscle Fiber	
	Slide 1	Slide 2
1. Glycerol alone	mm	mm
2. K^+/Mg^{2+} salt solution alone	mm	—
3. ATP alone	—	mm
4. Both ATP and salt solution	mm	mm

Conclusion: Muscle Fiber Contraction

- Was your hypothesis stated above supported? _____ Why or why not? _____

  _____

  _____

- To demonstrate that you understand the requirements for contraction, state the function of each of the substances listed in Table 12.2.

Table 12.2 Summary of Muscle Fiber Contraction

Substance	Function
Myosin	
Actin	
K^+/Mg^{2+} salt solution	
ATP	

Application for Daily Living

Bone Marrow Transplants

We are accustomed to thinking of a transplant as a procedure that replaces a nonfunctioning organ with one that works properly. A red bone marrow transplant doesn't quite work like that. Instead, red bone marrow is injected into the bloodstream and the recipient is receiving the cells that occur in the red bone marrow. The red bone marrow contains the precious stem cells capable of producing all the various cells in the blood. And they aren't ordinarily in the blood! However, these cells usually find their way home—the bones that ordinarily contain red bone marrow in adults, such as the sternum, breast bone, skull, hips, ribs, and spine.

The health reasons for needing a red bone marrow transplant are numerous, but chief among them are cancer patients whose own bone marrow was destroyed by treatment of cancer. As with any transplant, a careful match between donor and recipient is required.

Laboratory Review 12

Diaphysis **1.** Is compact bone located in the diaphysis or in the epiphyses?

Spongy **2.** Does compact bone or spongy bone contain red bone marrow?

Osteocytes **3.** What are bone cells called?

Cervical **4.** What are the vertebrae in the neck region called?

Femur **5.** Name the strongest bone in the lower limb.

Scapula/clavicle **6.** What bones are part of a pectoral girdle?

Ball in Socket
Adduction **7.** What type of joint movement occurs when a muscle moves a limb toward the midline of the body?

Rotation **8.** What type of joint movement occurs when a muscle moves a body part around its own axis?

Striation

ATP

Protein

Extend

Flex

Glutes

Hamstring

9. Skeletal muscle is voluntary, and its appearance is _____ because of the placement of actin and myosin filaments.

10. Glycerinated muscle requires the addition of what molecule to supply the energy for muscle contraction?

11. Actin and myosin are what type of biological molecule?

12. Does the quadriceps femoris flex or extend the leg?

13. Does the biceps brachii flex or extend the forearm?

14. What muscle forms the buttocks?

15. Name the muscle group antagonistic to the quadriceps femoris group.

Thought Questions

16. What bones protect the thoracic cavity?

    Ribs.....

17. When you see glycerinated muscle shorten, what is happening microscopically?

18. Release of a neurotransmitter at a neuromuscular junction causes a muscle fiber to contract. Label the neurotransmitter in this figure.

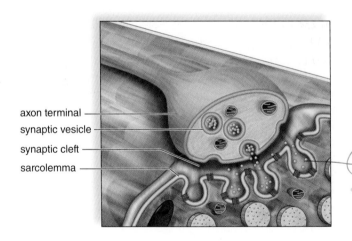

axon terminal
synaptic vesicle
synaptic cleft
sarcolemma

(green dots)
Ach
(neurotransmitter)

13

Nervous System and Senses

Learning Outcomes

13.1 Central Nervous System
- Identify the parts of the brain studied, and state the functions of each part.
- Give examples to show that the parts of the brain work together.
- Describe the anatomy of the spinal cord and tell how the cord functions as a relay station.

Prelab Question: Distinguish between the cerebral cortex and the cerebrum.

13.2 Peripheral Nervous System
- Distinguish between cranial nerves and spinal nerves on the basis of location and function.
- Describe the anatomy and physiology of a spinal reflex arc.

Prelab Question: How does the brain become aware that you have removed your hand from a hot stove?

13.3 The Human Eye
- Identify the parts of the eye and state a function for each part.

Prelab Question: What part of the eye contains the sensory receptors for sight?

13.4 The Human Ear
- Identify the parts of the ear and state a function for each part.

Prelab Question: What part of the ear contains the sensory receptors for hearing?

13.5 Sensory Receptors in Human Skin
- Describe the anatomy of the human skin and explain the distribution and function of sensory receptors.
- Relate the abundance of touch receptors to the ability to distinguish between two different touch points.

Prelab Question: What part of the skin contains sensory receptors?

13.6 Human Chemoreceptors
- Relate the ability to distinguish foods to the senses of smell and taste.

Prelab Question: Taste is dependent on what types of chemical stimuli?

Application for Daily Living: LASIK Surgery

Introduction

The nervous system has two major divisions: the central nervous system (CNS) consisting of the brain and spinal cord and the peripheral nervous system (PNS), which contains cranial nerves and spinal nerves (Fig. 13.1). Sensory receptors detect changes in environmental stimuli, and nerve impulses move along sensory nerve fibers to the brain and the spinal cord. The brain and spinal cord sum up the data before sending impulses via motor nerve fibers to effectors (muscles and glands) so a response to stimuli is possible. Nervous tissue consists of neurons; whereas the brain and spinal cord contain all parts of neurons, nerves contain only axons.

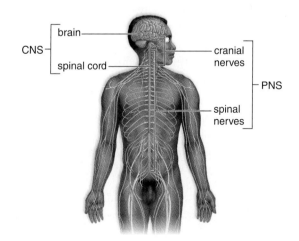

Figure 13.1 The nervous system.
The central nervous system (CNS) is in the midline of the body, and the peripheral nervous system (PNS) is outside the CNS.

13.1 Central Nervous System

The brain is the enlarged anterior end of the spinal cord; it contains centers that receive input from and can command other regions of the nervous system.

know Lobes

Preserved Sheep Brain

The sheep brain (Fig. 13.2) is often used to study the brain. It is easily available and large enough that individual parts can be identified.

Observation: Preserved Sheep Brain

Examine the exterior and a midsaggital (longitudinal) section of a preserved sheep brain or a model of the human brain, and with the help of Figure 13.2, identify the following.

1. **Ventricles:** interconnecting spaces that produce and serve as a reservoir for cerebrospinal fluid, which cushions the brain. Toward the anterior, note the lateral ventricle (on one longitudinal section) and similarly a lateral ventricle (on the other longitudinal section). Trace the second ventricle to the third and then the fourth ventricles.

2. **Cerebrum:** most developed area of the brain; responsible for higher mental capabilities. The cerebrum is divided into the right and left **cerebral hemispheres,** joined by the **corpus callosum,** a broad sheet of white matter. The outer portion of the cerebrum is highly convoluted and divided into the following surface lobes (see Fig. 13.4): *- largest overall structure*

   a. **Frontal lobe:** controls motor functions and permits voluntary muscle control; it also is responsible for abilities to think, problem solve, speak, and smell. *- Personality*

   b. **Parietal lobe:** receives information from sensory receptors located in the skin and also the taste receptors in the mouth. A groove called the **central sulcus** separates the frontal lobe from the parietal lobe. *- sensory*

   c. **Occipital lobe:** interprets visual input and combines visual images with other sensory experiences. The optic nerves split and enter opposite sides of the brain at the optic chiasma, located in the diencephalon.

   d. **Temporal lobe:** has sensory areas for hearing and smelling. The olfactory bulb contains nerve fibers that communicate with the olfactory cells in the nasal passages and take nerve impulses to the temporal lobe.

3. **Diencephalon:** portion of the brain where the third ventricle is located. The hypothalamus and thalamus are also located here. *- Contains Thalamus, hypothalamus, epithalamus*

   a. **Thalamus:** two connected lobes located in the roof of the third ventricle. The thalamus is the highest portion of the brain to receive sensory impulses before the cerebrum. It is believed to control which received impulses are passed on to the cerebrum. For this reason, the thalamus sometimes is called the "gatekeeper to the cerebrum." *- Relay station*

   b. **Hypothalamus:** forms the floor of the third ventricle and contains control centers for appetite, body temperature, and water balance. Its primary function is homeostasis. The hypothalamus also has centers for pleasure, reproductive behavior, hostility, and pain. *- Homeostasis*

4. **Cerebellum:** located just posterior to the cerebrum as you observe the brain dorsally, the cerebellum's two lobes make it appear rather like a butterfly. In cross section, the cerebellum has an internal pattern that looks like a tree. The cerebellum coordinates equilibrium and motor activity to produce smooth movements. *- movement/motor function -smoothing of (m) movement*

Bottom portion of Brain

Figure 13.2 The sheep brain.

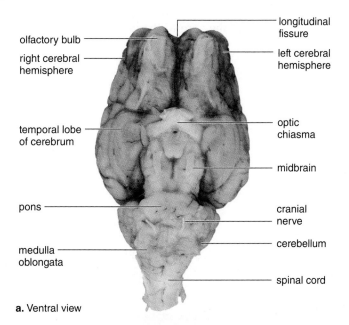

olfactory bulb

right cerebral hemisphere

temporal lobe of cerebrum

pons

medulla oblongata

longitudinal fissure

left cerebral hemisphere

optic chiasma

midbrain

cranial nerve

cerebellum

spinal cord

a. Ventral view

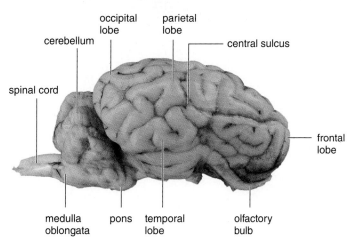

occipital lobe

parietal lobe

cerebellum

central sulcus

spinal cord

frontal lobe

medulla oblongata

pons

temporal lobe

olfactory bulb

b. Lateral view

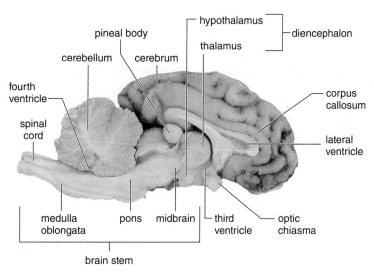

pineal body

hypothalamus

diencephalon

cerebellum

cerebrum

thalamus

fourth ventricle

corpus callosum

spinal cord

lateral ventricle

medulla oblongata

pons

midbrain

third ventricle

optic chiasma

brain stem

c. Longitudinal section

5. **Brain stem:** Part of the brain that connects with the spinal cord. Because it includes the pons and medulla oblongata, it contains centers for the functioning of internal organs; because of its location, it serves as a relay station for nerve impulses passing from the cord to the brain. Therefore, it helps keep the rest of the brain alert and functioning.

    a. **Midbrain:** anterior to the pons, the midbrain serves as a relay station for sensory input and motor output. It also contains a reflex center for eye muscles.

    b. **Pons:** the ventral, bulblike enlargement on the brain stem. It serves as a passageway for nerve impulses running between the medulla and the higher brain regions.

    c. **Medulla oblongata** (or simply **medulla**): the most posterior portion of the brain stem. It controls internal organs; for example, cardiac and breathing control centers are present in the medulla. Nerve impulses pass from the spinal cord through the medulla to higher brain regions.

The Human Brain

Label Figure 13.3 where indicated. Based on your knowledge of the sheep brain, complete Table 13.1 by stating the major functions of each part of the brain listed.

Table 13.1 Summary of Brain Functions	
Part	**Major Functions**
Cerebrum	
Cerebellum	
Diencephalon	
Thalamus	
Hypothalamus	
Brain stem	
Midbrain	
Pons	
Medulla oblongata	

Which parts of the brain would work together to achieve the following?

1. Good eye–hand coordination _____

2. Concentrating on homework when TV is playing _____

3. Avoiding dark alleys while walking home at night _____

4. Keeping the blood pressure constant _____

Figure 13.3   The human brain (longitudinal section).
The cerebrum is larger in humans than in sheep. Label this diagram where indicated.

Cerebral Lobes

As stated previously, the outer portion of the cerebrum is highly convoluted and divided into lobes as illustrated in Figure 13.4. The various sense organs send nerve impulses to a particular lobe where the nerve impulses are integrated to give us our senses of vision, hearing, smell, taste, and touch. Although not stated in Figure 13.4, the frontal lobe helps us remember smells of some significance to us. *Write the name of a sense next to the appropriate lobe in Figure 13.4.*

Figure 13.4   The cerebral lobes.
Each lobe has centers for integrating nerve impulses received from a particular type of sense organ. Our five senses result from this activity of the brain.

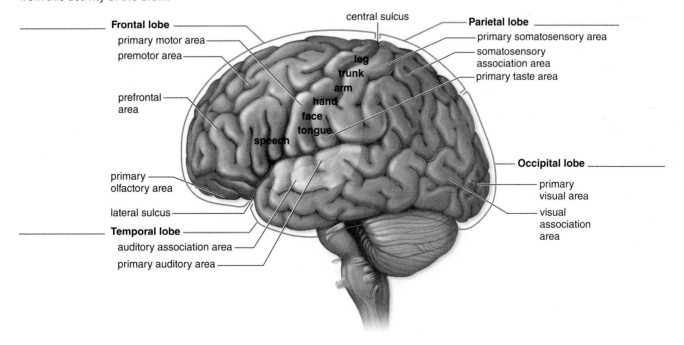

The Spinal Cord

The spinal cord is a part of the central nervous system. It lies in the middorsal region of the body and is protected by the vertebral column.

Observation: The Spinal Cord

1. Examine a prepared slide of a cross section of the spinal cord under the lowest magnification possible.
2. Identify the following with the help of Figure 13.5:
   a. **Gray matter:** a central, butterfly-shaped area composed of masses of short nerve fibers, interneurons, and motor neuron cell bodies
   b. **White matter:** masses of long fibers that lie outside the gray matter and carry impulses up and down the spinal cord. In living animals, white matter appears white because an insulating myelin sheath surrounds long fibers.

Figure 13.5   The spinal cord.
Photomicrograph of spinal cord cross section.

Know Anterior/post. op (handwritten)

— central canal *Broca nr* (handwritten)

— gray matter

— white matter

20×

13.2 Peripheral Nervous System   *Ventral = front* (handwritten)

The peripheral nervous system contains the cranial nerves and the spinal nerves. Twelve pairs of cranial nerves project from the inferior surface of the brain. The cranial nerves are largely concerned with nervous communication between the head, neck, and facial regions of the body and the brain. The 31 pairs of spinal nerves emerge from either side of the spinal cord (Fig. 13.6).

Spinal Nerves   *Post = sensory Ant = motor* (handwritten)

Each spinal nerve contains long fibers of sensory neurons and long fibers of motor neurons. In Figure 13.6, identify the following:

1. **Sensory neuron:** takes nerve impulses from a sensory receptor to the spinal cord. The cell body of a sensory neuron is in the dorsal root ganglion.
2. **Interneuron:** lies completely within the spinal cord. Some interneurons have long fibers and take nerve impulses to and from the brain. The neuron in Figure 13.6 transmits nerve impulses from the sensory neuron to the motor neuron.
3. **Motor neuron:** takes nerve impulses from the spinal cord to an effector—in this case, a muscle. Muscle contraction is one type of response to stimuli.

Suppose you were walking barefoot and stepped on a prickly sandbur. Describe the pathway of information, starting with the pain receptor in your foot, that would allow you to both feel and respond to this unwelcome stimulus. _____

Spinal Reflexes

A **reflex** is an involuntary and predictable response to a given stimulus that allows a quick response to environmental stimuli without communicating with the brain. In the spinal reflexes that follow, stretch receptors detect the tap, and sensory neurons conduct nerve impulses to interneurons in the spinal cord. The interneurons send a message via motor neurons to the effectors, muscles in the leg or foot. These reflexes are involuntary because the brain is not involved in formulating the response. Consciousness of the stimulus lags behind the response because information must be sent up the spinal cord to the brain before you can become aware of the tap.

Figure 13.6 Spinal nerves and spinal cord.

The arrows mark the path of nerve impulses from a sensory receptor to an effector. When a spinal reflex occurs, a sensory receptor is stimulated and generates nerve impulses that pass along the sensory neuron, the interneuron, and motor neuron causing the effector to respond.

Be able to label ★

Label Receptor
Control center
Sensory Neuron
Motor Neuron
Affector ★

pin

sensory receptor (in skin)

dorsal root ganglion ★

dorsal horn

central canal

white matter

gray matter

Dorsal

dorsal horn

cell body of sensory neuron ★

axon of sensory neuron

interneuron

effector (muscle)

axon of motor neuron

ventral root ★

cell body of motor neuron

ventral horn

ventral horn

Ventral

Ⓜ = Affector

Experimental Procedure: Spinal Reflex

One easily tested tendon reflex involves the **patellar tendon.** When this tendon is tapped with a reflex hammer (Fig. 13.7) or, in this experiment, with a meterstick, the attached muscle contracts and the leg rises. The stretch receptor generates nerve impulses which are transmitted along sensory neurons to the spinal cord. Nerve impulses from the cord then pass along motor neurons and stimulate the muscle, causing it to contract. Receptors in other tendons, such as the Achilles tendon, respond similarly.

Knee-Jerk (Patellar) Reflex

1. Have the subject sit on a table so that his or her legs hang freely.
2. Sharply tap one of the patellar tendons just below the patella (kneecap) with a meterstick.
3. In this relaxed state, does the leg flex (move toward the buttocks) or extend (move away from the buttocks)? _____

Figure 13.7 Knee-jerk reflex.

The quick response when the patellar tendon is stimulated by tapping with a reflex hammer indicates that a reflex has occurred.

Knee-jerk (patellar) reflex

13.3 The Human Eye

First you will observe an eye model and then perform two experimental procedures to test your own eyes.

Observation: The Human Eye

The eye is a special sense organ for detecting light rays in the environment. Examine a human eye model, and identify the structures listed in Table 13.2 and depicted in Figure 13.8. As you identify the structures state their functions.

Table 13.2	Parts of the Human Eye	
Part	**Location**	**Function**
Sclera	Outer layer of eye	Protects and supports eyeball
Cornea	Transparent portion of sclera	Refracts light rays
Choroid	Middle layer of eye	Absorbs stray light rays
Retina	Inner layer of eye	Contains receptors for sight
Rod cells (black and white vision)		
Cone cells (color vision)		
Fovea centralis	Special region of retina for cones	Makes acute vision possible
Lens	Between compartments	Refracts and focuses light rays
Ciliary body	Extension from choroid	Holds lens in place; functions in accommodation
Iris	Most anterior portion of choroid	Regulates light entrance
Pupil	Opening in middle of iris	Admits light
Aqueous and vitreous humors	Fluid media of eye	Transmit and refract light rays
Optic nerve	Extension from posterior of eye	Transmits impulses to occipital lobe of brain

(handwritten: → No Receptors → Blind spot)

(handwritten: optic nerve looks like tail)

Figure 13.8 Anatomy of the human eye.

(handwritten: Lens is like marble)

The Blind Spot of the Eye

The **blind spot** occurs where the optic nerve fibers exit the retina. No vision is possible at this location because of the absence of rod cells and cone cells.

Experimental Procedure: Blind Spot of the Eye

This Experimental Procedure requires a laboratory partner. Figure 13.9 shows a small circle and a cross several centimeters apart.

Figure 13.9   Blind spot.
This dark circle (or cross) will disappear at one location because there are no rod cells or cone cells at each eye's blind spot, where vision does not occur.

Left Eye

1. Hold Figure 13.9 approximately 30 cm from your eyes. The cross should be directly in front of your left eye. If you wear glasses, keep them on.
2. Close your right eye.
3. Stare only at the cross with your left eye. You should also be able to see the circle in the same field of vision. Slowly move the paper toward you until the circle disappears.
4. Repeat the procedure as many times as needed to find the blind spot.
5. Then slowly move the paper closer to your eyes until the circle reappears. Because only your left eye is open, you have found the blind spot of your left eye.
6. With your partner's help, measure the distance from your eye to the paper when the circle first

   disappeared. Left eye: _____ cm

Right Eye

1. Hold Figure 13.9 approximately 30 cm from your eyes. The circle should be directly in front of your right eye. If you wear glasses, keep them on.
2. Close your left eye.
3. Stare only at the circle with your right eye. You should also be able to see the cross in the same field of vision. Slowly move the paper toward you until the cross disappears.
4. Repeat the procedure as many times as needed to find the blind spot.
5. Then slowly move the paper closer to your eyes until the cross reappears. Because only your right eye is open, you have found the blind spot of your right eye.
6. With your partner's help, measure the distance from your eye to the paper when the cross first

   disappeared. Right eye: _____ cm

In this exercise, you created an artificial situation in which you became aware of how your perception of the world can be constrained by the eye's anatomy. Although the eye detects patterns of light and color, it is the brain that determines what we visually perceive. The brain can fill in data that is missing based in part on past experiences.

Accommodation of the Eye *(1) In eye begin to weaken w/ age*

When the eye accommodates to see objects at different distances, the shape of the lens changes. The lens shape is controlled by the ciliary muscles attached to it. When you are looking at a distant object, the lens is in a flattened state. When you are looking at a closer object, the lens becomes more rounded. The elasticity of the lens determines how well the eye can accommodate. Lens elasticity decreases with increasing age, a condition called **presbyopia.** Presbyopia is the reason many older people need bifocals to see near objects.

Experimental Procedure: Accommodation of the Eye

This Experimental Procedure requires a laboratory partner. It tests accommodation of either your left or right eye.

1. Hold a pencil upright by the eraser and at arm's length in front of whichever of your eyes you are testing (Fig. 13.10).
2. Close the opposite eye.
3. Move the pencil from arm's length toward your eye.
4. Focus on the end of the pencil.
5. Move the pencil toward you until the end is out of focus. Measure the distance (in centimeters)

   between the pencil and your eye: _____ cm

6. At what distance can your eye no longer

   accommodate for distance? _____ cm

7. If you wear glasses, repeat this experiment without your glasses, and note the accommodation distance of your eye without glasses:

   _____ cm. (Contact lens wearers need not make these determinations, and they should write the words *contact lens* in this blank.)

8. The "younger" lens can easily accommodate for closer distances. The nearest point at which the end of the pencil can be clearly seen is called the **near point.** The more elastic the lens,

   the "younger" the eye (Table 13.3). How "old" is the eye you tested? _____

Figure 13.10   Accommodation.
When testing the ability of your eyes to accommodate to see a near object, always keep the pencil in this position.

Table 13.3 Near Point and Age Correlation						
Age (years)	10	20	30	40	50	60
Near point (cm)	9	10	13	18	50	83

13.4 The Human Ear

The human ear, whose parts are listed and depicted in Table 13.4 and Figure 13.11, serves two functions: hearing and balance.

Observation: The Human Ear

Examine a human ear model, and find the structures depicted in Figure 13.11 and listed in Table 13.4.

Table 13.4 Parts of the Human Ear

Part	Medium	Function	Mechanoreceptor
Outer ear	Air		
Pinna		Collects sound waves	—
Auditory canal		Filters air	—
Middle ear	Air		
Tympanic membrane and ossicles		Amplify sound waves	—
Auditory tube		Equalizes air pressure	—
Inner ear	Fluid		
Semicircular canals		Rotational equilibrium	Stereocilia embedded in cupula
Vestibule (contains utricle and saccule)		Gravitational equilibrium	Stereocilia embedded in otolithic membrane
Cochlea (spiral organ)		Hearing	Stereocilia embedded in tectorial membrane

Figure 13.11 Anatomy of the human ear.
The outer ear extends from the pinna to the tympanic membrane. The middle ear extends from the tympanic membrane to the oval window. The inner ear encompasses the semicircular canals, the vestibule, and the cochlea.

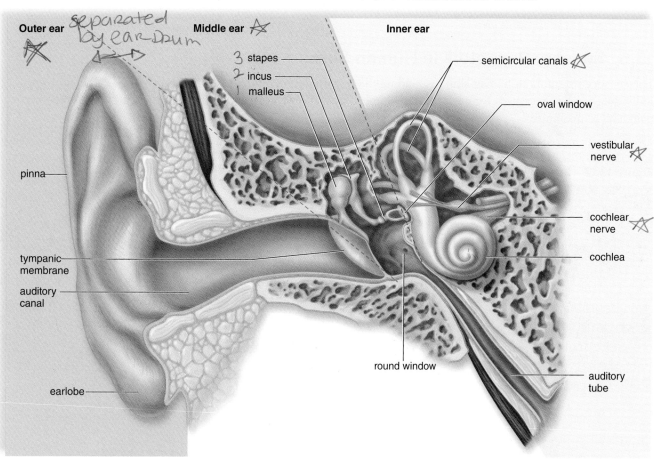

Physiology of the Human Ear When you hear, sound waves are picked up by the **tympanic membrane** and amplified by the **malleus, incus,** and **stapes.** This creates pressure waves in the canals of the **cochlea** that lead to stimulation of **hair cells,** the receptors for hearing. Hair cells in the utricle and saccule of the vestibule and in semicircular canals are receptors for equilibrium (i.e., balance). Nerve impulses from the sensory receptors in the ears travel by way of the cochlear nerve and the vestibular nerve to the brain and eventually are interpreted by the _____ lobe of the brain.

Humans locate the direction of sound according to how fast it is detected by either or both ears. A difference in the hearing ability of the two ears can lead to a mistaken judgment about the direction of sound. You and a laboratory partner should perform this Experimental Procedure on each other. Enter the data for your ears, not your partner's ears, in the spaces provided.

1. Ask the subject to be seated, with eyes closed. Then strike a tuning fork or rap two spoons together at the five locations listed in number 2. Use a random order.
2. Ask the subject to give the exact location of the sound in relation to his or her head. Record the subject's perceptions when the sound is

    a. Directly below and behind the head _____

    b. Directly behind the head _____

    c. Directly above the head _____

    d. Directly in front of the face _____

    e. To the side of the head _____

3. Is there an apparent difference in hearing between your two ears? _____

13.5 Sensory Receptors in Human Skin

The sensory receptors in human skin respond to touch, pain, temperature, and pressure (Fig. 13.12). There are individual sensory receptors for each of these stimuli, as well as free nerve endings able to respond to pressure, pain, and temperature.

Figure 13.12 Sensory receptors in the skin.
Each type of receptor shown responds primarily to a particular stimulus.

Sense of Touch

The dermis of the skin contains touch receptors, whose concentration differs in various parts of the body.

Experimental Procedure: Sense of Touch

You will need a laboratory partner to perform this Experimental Procedure. Enter your data, not the data of your partner, in the spaces provided.

1. Ask the subject to be seated, with eyes closed.
2. Then test the subject's ability to discriminate between the two points of a hairpin or a pair of scissors at the four locations listed in number 5.
3. Hold the points of the hairpin or scissors on the given skin area, with both of the points simultaneously and gently touching the subject.
4. Ask the subject whether the experience involves one or two touch sensations.
5. Record the shortest distance between the hairpin or scissor points for a two-point discrimination.

   a. Forearm: _____ mm

   b. Back of the neck: _____ mm

   c. Index finger: _____ mm

   d. Back of the hand: _____ mm

6. Which of these areas apparently contains the greatest density of touch receptors? _____

   Why is this useful? _____

7. Do you have a sense of touch at every point in your skin? _____ Explain. _____

   _____

8. What specific part of the brain processes nerve impulses from touch and pain receptors? See Figure 13.4.

   _____

Sense of Heat and Cold

Temperature receptors respond to a change in temperature.

Experimental Procedure: Sense of Heat and Cold

1. Obtain three 1,000 mL beakers, and fill one with ice water, one with tap water at room temperature, and one with warm water (45°–50°C).
2. Immerse your left hand in the ice-water beaker and your right hand in the warm-water beaker for 30 seconds.
3. Then place both hands in the beaker with room-temperature tap water.
4. Record the sensation in the right and left hands.

   a. Right hand: _____

   b. Left hand: _____

5. Explain your results. _____

   _____

13.6 Human Chemoreceptors

The taste receptors, called _____, located in the mouth, and the smell receptors, called _____, located in the nasal cavities, are the chemoreceptors that respond to molecules in the air and water. Nerve impulses from taste receptors go to the _____ lobe of the brain while those from smell receptors go to the _____ lobe of the brain.

Experimental Procedure: Sense of Taste and Smell

You will need a laboratory partner to perform the following procedures. It will not be necessary for all tests to be performed on both partners. You should take turns being either the subject or the experimenter. Dispose of used cotton swabs in a hazardous waste container or as directed by your instructor.

1. Students work in groups. Each group has one experimenter and several subjects.
2. The experimenter should obtain a LifeSavers candy from the various flavors available, without letting the subject know what flavor it is.
3. The subject closes both eyes and holds his or her nose.
4. The experimenter gives the LifeSavers candy to the subject, who places it on his or her tongue.
5. The subject, while still holding his or her nose, guesses the flavor of the candy. The experimenter records the guess in Table 13.5.
6. The subject releases his or her nose and guesses the flavor again. The experimenter records the guess and the actual flavor in Table 13.5.

Table 13.5   Taste and Smell Experiment

Subject	Actual Flavor	Flavor While Holding Nose	Flavor After Releasing Nose
1			
2			
3			
4			
5			

Conclusions: Sense of Taste and Smell

- From your results, how would you say that smell affects the taste of LifeSavers candy?

  _____

- What do you conclude about the effect of smell on your sense of taste?

  _____

LASIK Surgery

The focusing ability of the lens to place an image on the retina so we can see is assisted by the cornea. The cornea refracts or bends the light rays, and then the lens takes on the chore from then on. This accounts for why LASIK eye surgery, which reshapes the cornea, works. The traditional LASIK vision correction involves two steps: (1) First, the surgeon has to make a flap in the outer surface of the eye to expose the underlying cornea. (2) Then, the cornea is reshaped.

Anyone considering undergoing LASIK eye surgery should be aware that various side effects have been seen from dry eyes to severe glare when driving at night. Therefore, the procedure should be discussed with a physician or optician first. They can recommend a reliable clinic and surgeon and will also be able to advise whether there is any reason why LASIK eye surgery might not work for you.

Laboratory Review 13

Cerebrum	1.	What part of the brain is largest in humans?
Cerebellum	2.	What part of the brain controls muscular coordination?
Pia mater	3.	What is the most inferior portion of the brain stem?
vertebral column	4.	What structures protect the spinal cord?
Gray matter	5.	Are motor neuron cell bodies located in the gray or white matter of the spinal cord?
motor	6.	What type of neuron is found completely within the central nervous system?
Sensory	7.	Which neuron's cell body is in the dorsal root ganglion?
Retina	8.	What part of the eye contains the sensory receptors for sight?
Optic Nerve	9.	Where on the retina is the blind spot located?
Sclera	10.	What do you call the outer layer of the eye?
Cochlea	11.	What part of the ear contains the sensory receptors for hearing?
middle ear	12.	In which part of the ear are the malleus, incus, and stapes located?
Dermis	13.	What layer of the skin contains sensory receptors for touch?
unevenly	14.	Are touch receptors distributed evenly or unevenly in the skin?
	15.	What senses are dependent on chemoreceptors?
Umami	16.	The four taste sensations are sour, salty, bitter, and _____.
	17.	What advantages are associated with the spinal cord and spinal nerves functioning below the level of consciousness?
	18.	Identify the type of neuron responsible for transmitting nerve impulses from the spinal cord to an effector.

Thought Questions

19. Trace the path of light in the human eye through each structure or compartment—from the exterior to the retina. How do nerve impulses from the retina reach the occipital lobe of the brain?

20. Trace the path of sound waves in the human ear—from the tympanic membrane to the sensory receptors for hearing.

|BIOLOGY

Find even more resources to learn the chapter concepts including animations, tutorial videos, and interactive practice questions.

McGraw-Hill Access Science Website

An online encyclopedia of science and technology that provides information, including videos, that can enhance the laboratory experience.

www.accessscience.com

The following LearnSmart Lab contains exercises that are related to the content of this chapter:

• The Nervous System and Senses

14

Reproduction and Development

Learning Outcomes

14.1 Male and Female Reproductive Systems
- Trace the path of sperm in the male and the egg in the female.
- Locate on a model and state a function for each major organ in the male and female reproductive systems.
- Identify a slide of the testis and the ovary and the major structures in these organs.
- Describe the specialization of a sperm and of an egg for their functions.

Prelab Question: What is the most significant biological contribution of a male and a female to an offspring?

14.2 Embryonic Development
- Identify the cellular stages of development with reference to slides of early sea star development.
- Identify the tissue stages of development with reference to slides of frog development.
- Associate the germ layers with the development of various organs.
- Identify which organs develop first in a vertebrate embryo (e.g., frog, chick, and human).
- When presented with a sequence of human embryos, point out and discuss aspects of their increasing complexity.

Prelab Question: Which human organs develop first during development?

14.3 Extraembryonic Membranes, the Placenta, and the Umbilical Cord
- Distinguish between and give a function for the extraembryonic membranes, the placenta, and the umbilical cord.
- Trace the development of the extraembryonic membranes during embryonic development, and state a function for each membrane.

Prelab Question: Which extraembryonic membrane becomes the placenta?

14.4 Fetal Development
- Trace the main events of human fetal development.

Prelab Question: Account for why the respiratory system is not functional until the sixth month or later.

***Application for Daily Living:* Cord Around Baby's Neck**

Introduction

The sperm and egg contribute chromosomes to the offspring. The testes in males and the ovaries in females produce the gametes and also the hormones that maintain the organs and the characteristics we associate with a male and female. In humans, embryonic development begins when the sperm fertilizes the egg and a zygote is formed. Cell division produces a ball of cells that arrange themselves into three layers, called the **germ layers.** It is possible to associate the development of particular organs with a specific germ layer. Embryonic development comes to a close when all the basic organs have formed. Refinements and an increase in size occur during fetal development.

zygote

embryo at one week; implants in uterine wall

embryo at eight weeks

fetus at three months

fetus at five months

14.1 Male and Female Reproductive Systems

Only the urinary and reproductive systems have a different anatomy and physiology in males and females.

Observation: Male Reproductive System Model

The functions of the male reproductive system are to (1) produce sperm and the male sex hormones within the testes, (2) transport the sperm in ducts until they exit through the penis, (3) nourish and provide a medium for sperm in which they can survive, and (4) deliver sperm to the vagina of the female. Identify the following organs in a model (Fig. 14.1):

1. **Testes.** The primary sex organs of males where the sperm and male sex hormones (e.g., testosterone) are produced. The testes fulfill the first function noted previously for the male reproductive system.
2. **Epididymis.** A tightly coiled, threadlike tube where sperm are stored until they are mature and capable of fertilizing an egg.
3. **Vas deferens.** A muscular tube that conducts sperm to the ejaculatory duct.
4. **Ejaculatory duct.** When the vas deferens and a duct from the seminal vesicle meet, they form a short ejaculatory duct, which passes through the prostate gland and empties into the urethra.
5. **Urethra.** The male urethra is located in the penis. The urethra either conducts semen or it conducts urine to the exterior and does not pass them both at the same time. What function of the male reproductive system previously listed is fulfilled by the ducts (2–5) just described? _____
6. **Seminal vesicles,** the **prostate gland,** and the **bulbourethral glands.** These glands do not produce hormones, but they do produce secretions that account for why semen (the substance that exits the penis during ejaculation) is a thick fluid. The secretions of these glands are favorable to the health of the sperm. What function noted previously for the male reproductive system is fulfilled by these glands? _____

The prostate gland frequently enlarges in older men and blocks the urethra. Under these conditions, a male is apt to have difficulty urinating. Explain. _____

Ductus Deferens = Vas Deferens

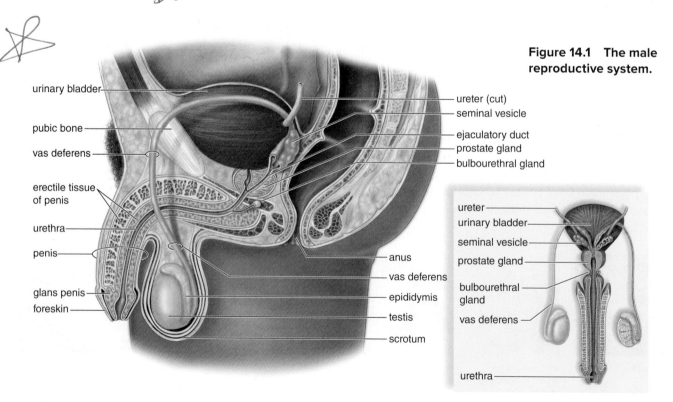

Figure 14.1   The male reproductive system.

urinary bladder
pubic bone
vas deferens
erectile tissue of penis
urethra
penis
glans penis
foreskin

ureter (cut)
seminal vesicle
ejaculatory duct
prostate gland
bulbourethral gland

anus
vas deferens
epididymis
testis
scrotum

ureter
urinary bladder
seminal vesicle
prostate gland
bulbourethral gland
vas deferens

urethra

Testes

1. Obtain a slide of the testis. As shown in Figure 14.2*a,* a testis (sing.) contains many seminiferous tubules where spermatogenesis (production of sperm) occurs. Note under low power the circular nature of a tubule.

2. Switch to high power, and observe one tubule in particular. Between the tubules, try to identify interstitial cells, which produce testosterone. Testosterone enters the body by way of the blood and not by way of ducts. Explain why a vasectomy, a cutting of the vas deferens, causes a male to be sterile but has no effect on his masculinity. _____

_____

3. **Spermatogenesis** occurs in the seminiferous tubules (Fig. 14.2*b*). Identify:

   Sertoli cells nuclei. The cytoplasm of a Sertoli cell surrounds a cell undergoing spermatogenesis.

   Germ cells. First spermatogonia, next spermatocytes, and finally spermatids occur, but they are hard to identify. Males produce four viable spermatids for each spermatogenesis. Each spermatid becomes a sperm.

   Sperm. Sperm tails look like thin, fine, dark lines in the lumen of the tubule. The sperm are fully developed but immature. They will complete their maturation in the epididymis. Why is it important for sperm cells to have a tail? _____

_____

Sperm

Obtain a prepared slide of sperm and compare what you see to Figure 14.2*c*. An acrosome contains enzymes that digest a path through cells that surround an egg and the outer covering of an egg so that one sperm can enter.

State another way sperm are specialized. _____

The ejaculated semen of a normal male contains several hundred million sperm, each a product of spermatogenesis, but only one sperm normally enters an egg. Speculate why so many sperm are needed for fertilization to occur. _____

_____

a. testis
(cut to show lobules)

b. seminiferous tubules   100×   **c.** sperm

Figure 14.2   Testis and sperm.

Observation: Female Reproductive System Model

The function of the female reproductive system (Fig. 14.3) is to (1) produce eggs and sex hormones within the ovaries; (2) transport eggs from the ovaries to the uterus; (3) receive the sperm in the vagina, also the birth canal; and (4) protect the fertilized egg (zygote) until it matures and is born. Identify the following organs in a model (see Fig. 14.3):

1. **Ovaries.** The primary sex organs of females where eggs and female sex hormones (e.g., estrogen and progesterone) are produced.
2. **Oviducts.** At the end nearest the ovaries, the oviducts have fimbriae that sweep over an ovary so that a released egg enters the oviduct. Cilia that line the oviduct and muscular contraction of its walls propel the egg toward the uterus. Fertilization, if it occurs, happens in an oviduct.
3. **Uterus.** A thick-walled, muscular organ about the size and shape of an inverted pear, where development of the embryo and fetus occur. The cervix surrounds the opening to the uterus.
4. **Vagina.** The uterus empties into the vagina, which makes a 45° angle with the small of the back.

✗ oviduct = Fallopian tube

Figure 14.3   The female reproductive system.

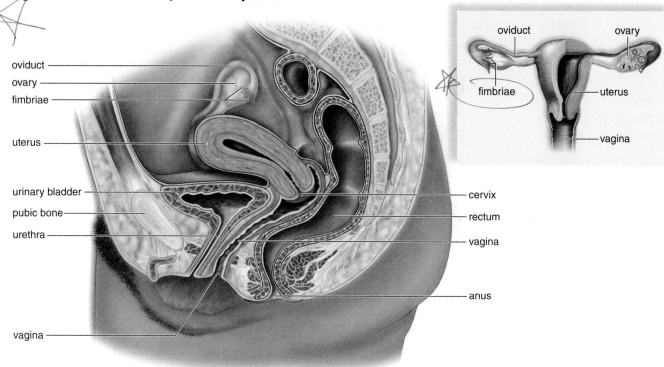

Observation: Microscopic Examination of the Ovary

Ovary

1. Examine a prepared slide of an ovary, and refer to Figure 14.4 for help in identifying the structures.
2. Locate a **primary follicle,** which appears as a circle of small cells surrounding a somewhat larger cell, the **primary oocyte.**
3. Find a **secondary follicle,** and switch to high power. Note the **secondary oocyte** (egg), surrounded by numerous cells, to one side of the liquid-filled follicle. Why is it important for females to produce an egg that has plentiful cytoplasm? _____

4. Look for a large, fluid-filled **vesicular (Graafian) follicle,** which contains a mature secondary oocyte to one side. This follicle will be next to the outer surface of the ovary because it is the type of follicle that releases the egg during ovulation.

5. Look for the remains of the **corpus luteum,** which will look like scar tissue. The corpus luteum develops after the vesicular follicle has released its egg, and then later it deteriorates. Not all slides will contain a corpus luteum and a vesicular follicle because they may not have been present when the slide was made.

6. **Oogenesis**
   The primary follicle and the secondary follicle are the site of oogenesis (production of an egg). Each oogenesis event in females produces a single egg. In contrast to males, females produce one egg a month.

Figure 14.4   Ovary and oocytes.

Summary of Reproductive Systems

- Complete Table 14.1 to describe the differences between the male and female human reproductive systems. A **gonad** is an organ that produces gametes, the sperm and egg.

Table 14.1 Comparison of Human Male and Female Reproductive Systems		
	Male	**Female**
Gonad		
Duct from gonad		
Duct from gonad leads to		
Copulatory organ		

14.2 Embryonic Development

We will divide embryonic development into three stages: cellular, tissue layer, and organ development. In human beings, it takes two months to complete embryonic development. It is impossible for us to view the stages of embryonic development in a human being, so we will use the sea star, frog, and chick as our observational material.

Cellular Stages of Development

The cellular stages of development include the following:

- **Zygote formation:** A single sperm fertilizes an egg and the result is a zygote, the first cell of the new individual.
- **Morula formation:** The zygote divides into a number of smaller cells until there is a cluster of 16–32 cells called a morula.
- **Blastula formation:** The morula becomes a blastula, a hollow ball of cells.

Observation: Cellular Stages of Development in the Sea Star

The cellular stages of development are remarkably similar in all animals. Therefore, we can view slides of sea star development to study the cellular stages of human development (Fig. 14.5). A sea star is an invertebrate that develops in the ocean and, therefore, will develop easily in the laboratory where it can be observed.

Obtain slides or view a model of sea star development and note the following:

1. **Zygote.** Both plants and animals begin life as a single cell, a zygote. A zygote contains chromosomes from each parent. Explain. _____

_____

2. **Cleavage.** View slides showing various numbers of cells due to the process of cleavage, cell division without growth until the morula stage. Is the morula about the same size as the zygote? _____

   Explain. _____

_____

3. **Blastula.** The cavity of a blastula is called the blastocoel. _Label blastocoel in Figure 14.5._ The formation of a hollow cavity is important to the next stage of development.

Figure 14.5 Starfish development.
All animals, including starfish and humans, go through the same cellular stages from cleavage to blastula.

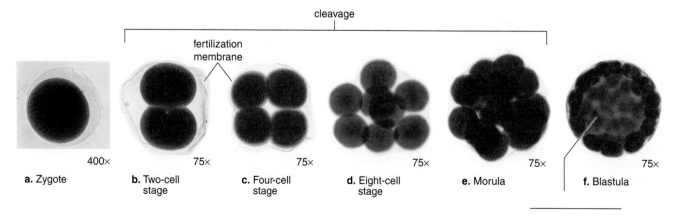

Observation: Cellular Stages of Development in Humans

In Figure 14.6, or in a model of human development, observe the same stages of development already observed in sea star slides. Also, observe that fertilization in humans occurs in an oviduct following ovulation. As the embryo undergoes cleavage, it travels in the oviduct to the uterus.

If the embryo splits at the 2-cell stage, the result is identical twins. (Fraternal twins arise when two separate eggs are fertilized.) How might you account for the development of identical triplets?

_____

_____

The blastula in humans is called a blastocyst. The blastocyst contains an **inner cell mass** that becomes the embryo, and the outer group of cells (the trophoblast) will become membranes that nourish and protect it. At about day 6, the blastocyst has reached the uterus and implants into the uterine wall, where it will receive nourishment from the mother's bloodstream.

What's the main difference between the cellular stages in a sea star and in a human?

_____

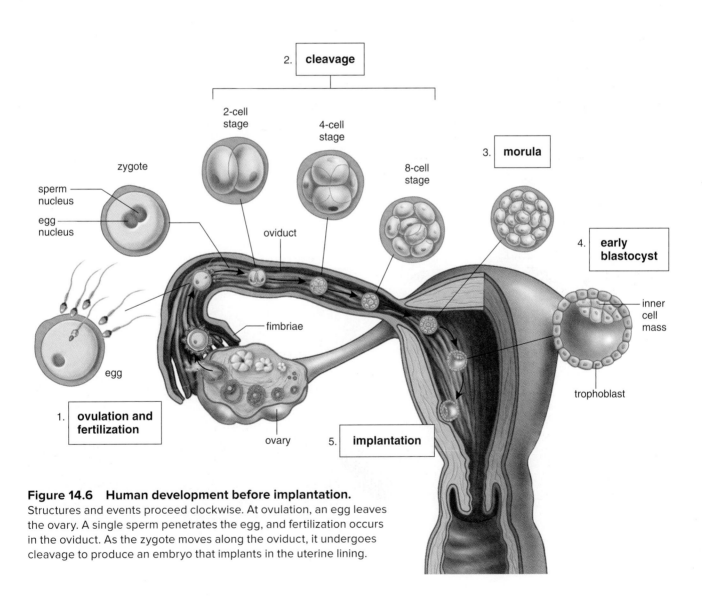

Figure 14.6 Human development before implantation.
Structures and events proceed clockwise. At ovulation, an egg leaves the ovary. A single sperm penetrates the egg, and fertilization occurs in the oviduct. As the zygote moves along the oviduct, it undergoes cleavage to produce an embryo that implants in the uterine lining.

Tissue Stages of Development

The tissue stages of development include the following:

- **Early gastrula stage.** This stage begins when certain cells begin to push or invaginate into the blastocoel, creating a double layer of cells. The outer layer is called the ectoderm, and the inner layer is called the endoderm.
- **Late gastrula stage.** Gastrulation is not complete until there are three layers of cells. The third layer called mesoderm occurs between the other two layers already mentioned.

Observation: Tissue Stages of Development in a Frog

It is traditional to view frog gastrulation. A frog is a vertebrate, and so its development is expected to be closer to that of a human than is a sea star. In Figure 14.7, note that the yellow (vegetal pole) cells are heavily laden with yolk, and the blue (animal pole) cells are the ones that invaginate into the blastocoel forming the early gastrula.

1. **Early gastrula stage.** Obtain a cross section of a frog gastrula. Most likely, your slide is the equivalent of Figure 14.7*b*, number 3, in which case you will see two cavities, the old blastocoel and newly forming *archenteron*, which forms once the animal pole cells have invaginated. The archenteron will become the digestive tract.
2. **Late gastrula stage.** In Figure 14.7, note that a third layer of cells, the mesoderm, is colored red and that it develops between the ectoderm and endoderm.

Figure 14.7   Drawings of frog developmental stages.
a. During cleavage, the number of cells increases but overall size remains the same. **b.** During gastrulation, three tissue layers form. **c.** During neurulation, the notochord and neural tube form.

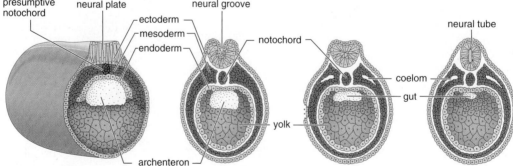

In a model of human development, observe the same stages of development already observed in frog slides. After implantation, gastrulation in humans turns the inner cell mass into the **embryonic disk.** Figure 14.8 shows the embryonic disk, which has the three layers of cells we have been discussing: the ectoderm, mesoderm, and endoderm. Figure 14.8 also shows the significance of these layers, often called the **germ layers.** The future organs of an individual can be traced back to one of the germ layers.

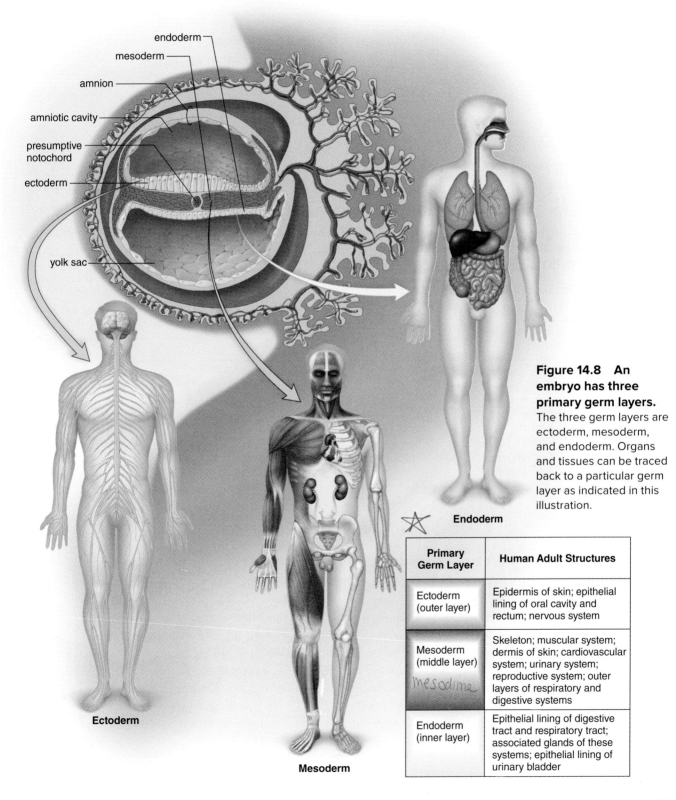

Figure 14.8 An embryo has three primary germ layers. The three germ layers are ectoderm, mesoderm, and endoderm. Organs and tissues can be traced back to a particular germ layer as indicated in this illustration.

Primary Germ Layer	Human Adult Structures
Ectoderm (outer layer)	Epidermis of skin; epithelial lining of oral cavity and rectum; nervous system
Mesoderm (middle layer)	Skeleton; muscular system; dermis of skin; cardiovascular system; urinary system; reproductive system; outer layers of respiratory and digestive systems
Endoderm (inner layer)	Epithelial lining of digestive tract and respiratory tract; associated glands of these systems; epithelial lining of urinary bladder

Organ Stages of Development

As soon as all three embryonic tissue layers (ectoderm, endoderm and mesoderm) are established, the organ level of development begins. It continues until development is complete. The first organs to develop are the

- Digestive tract. You have already observed the start of the archenteron during gastrulation.
- Spinal cord and brain
- Heart

Observation: Development of the Spinal Cord and Brain

One of the first systems to form is the nervous system. Why might it be beneficial for the nervous system to begin development first? _____ _____

_____

1. Obtain a cross section of a frog neurula stage, and match it to one of the drawings in Figure 14.7c.

   Which drawing seems to best match your slide? _____

   Your instructor will confirm your match for you.

2. A neural tube develops from ectoderm (Fig. 14.7c). Can you see how? _____

   When neural folds rise up and fuse, the neural tube has formed. The neural tube, which runs the length of the embryo, is the first sign of the central nervous system. The nerve cord, also called the spinal cord, and the brain both develop from the neural tube.

   Notice how the neural tube develops above the notochord, a dorsal supporting rod that later becomes the vertebral column. Why would you expect the neural tube, which becomes the spinal cord, to develop in the same vicinity as the notochord, which becomes the vertebral column? _____

_____

Observation: Development of the Heart

A chick embryo offers an opportunity to view a beating heart in an embryo. Your instructor may show you various stages. In particular you will want to observe the 48-hour chick embryo.

Observing Live Chick Embryos

Use the following procedure for selecting and opening the eggs of live chick embryos:

1. Choose an egg of the proper age to remove from the incubator, and put a penciled × on the uppermost side. The embryo is just below the shell.
2. Add warmed chicken Ringer solution to a finger bowl until the bowl is about half full. (Chicken Ringer solution is an isotonic salt solution for chick tissue that maintains the living state.) The chicken Ringer solution should not cover the yolk of the egg.
3. On the edge of the dish, gently crack the egg on the side opposite the ×.
4. With your thumbs placed over the ×, hold the egg in the chicken Ringer solution while you pry it open from below and allow its contents to enter the solution. If you open the egg too slowly or too quickly, the shell may damage the delicate membranes surrounding the embryo.

1. Follow the standard procedure (see page 192) for selecting and opening an egg containing a 48-hour chick embryo.
2. The embryo has turned so that the head region is lying on its side. Refer to Figure 14.9, and identify the following:

   a. **Shape of the embryo,** which has started to bend. The head is now almost touching the heart.

   b. **Heart,** contracting and circulating blood. Can you make out a ventricle, an atrium, and the aortic arches in the region below the head? Later, only one aortic arch will remain.

   c. **Vitelline arteries** and **veins,** which extend over the yolk. The vitelline veins carry nutrients from the yolk sac to the embryo.

   d. **Brain** with several distinct regions.

   e. **Eye,** which has a developing lens.

   f. **Margin (edge) of the amnion,** which can be seen above the vitelline arteries (see next section for amnion).

   g. **Somites,** blocks of developing muscle tissue that differentiate from mesoderm, which now number 24 pairs.

   h. **Caudal fold** of the amnion. The embryo will be completely enveloped when the head fold and caudal fold meet the margin of the amnion.

   i. **Neural tube,** which runs the length of the embryo, is the first sign of the central nervous system.

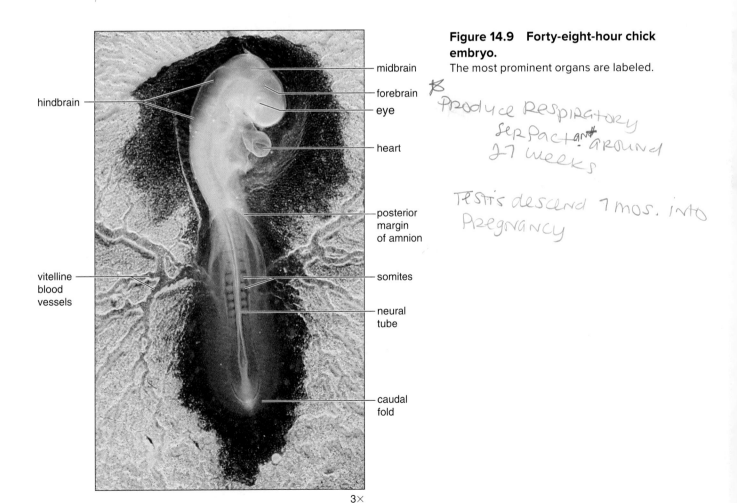

Figure 14.9 Forty-eight-hour chick embryo.
The most prominent organs are labeled.

produce Respiratory surpactant around 27 weeks

Testis descend 7 mos. into pregnancy

3×

Study models or other study aids available that show the development of the nervous system and the heart in human beings and/or show models of human embryos of different ages. Also view Figure 14.10, which depicts the external appearance of the embryo from the fourth to the seventh week of development.

During the embryonic period of development, the growing baby is susceptible to environmental influences, including the following:

- Drugs, such as alcohol; certain prescriptions; and recreational drugs. These can cause birth defects.
- Infections such as rubella, also called German measles, and other viral infections.
- Nutritional deficiencies.
- X-rays or radiation therapy.

Figure 14.10 External appearance of the embryo.
a. Weeks 4 to 5. **b.** Weeks 6 to 7.

a. Weeks 4 and 5
- Head dominant, but body getting longer.
- Limb buds are visible.
- Eyes and ears begin to form.
- Tissue for vertebrae extend into tail.

b. Weeks 6 and 7
- Head still dominant, but tail has disappeared.
- Facial features continue to develop.
- Hands and feet have digits.
- All organs more developed.

14.3 Extraembryonic Membranes, the Placenta, and the Umbilical Cord

- The **extraembryonic membranes** take their name from the observation that they are not part of the embryo proper. They are outside the embryo, and therefore they are "extra."
- The **placenta** is the structure that provides the embryo with nutrient molecules and oxygen and takes away its waste molecules, such as carbon dioxide. The fetal half of the placenta is the chorionic villi, which contain fetal capillaries. The maternal half of the placenta is capillaries in the uterine wall.
- The **umbilical cord** is a tubular structure that contains two of the extraembryonic membranes (the allantois and the yolk sac) and also the umbilical blood vessels. The umbilical blood vessels bring fetal blood to and from the placenta. When a baby is born and begins to breathe on its own, the umbilical cord is cut and the remnants become the navel.

In this drawing, *label the umbilical cord,* which contains the umbilical blood vessels. Also *label the placenta,* which contains the maternal blood vessels.

c. _____

a. _____

b. _____

maternal blood vessels

Observation: The Extraembryonic Membranes

In a model, and in Figure 14.11, trace the development of the extraembryonic membranes. Also, note the development of the placenta and the umbilical cord. The extraembryonic membranes are the

- **Chorion.** The chorion begins to form at the blastocyst stage of development. The outer layer of cells surrounding the inner cell mass of the blastocyst becomes the chorion. Notice in Figures 14.8 and 14.11, the treelike **chorionic villi** are a part of the chorion that will become the placenta.
- **Amnion.** Forms the amniotic cavity, which envelops the fetus and contains the amniotic fluid that cushions and protects the fetus (Fig. 14.12). All animals, whether the sea star, the frog, the chick, or a human, develop in an aqueous environment. Birth of a human is imminent when "the water breaks," the loss of the amniotic fluid.
- **Allantois.** The allantois extends into the umbilical cord. It accumulates the small amount of urine produced by the fetal kidneys and later gives rise to the urinary bladder. Its blood vessels become the umbilical blood vessels.
- **Yolk sac.** The yolk sac is the first embryonic membrane to appear. In the chick, the yolk sac does contain yolk, food for the developing embryo. In humans, the yolk sac contains plentiful blood vessels and is the first site of blood cell formation.

Figure 14.11 Development of extraembryonic membranes.
a. At first, no organs are present in the embryo, only tissues. The amniotic cavity is above the embryonic disk, and the yolk sac is below. The chorionic villi are present. **b, c.** The allantois and yolk sac, two more extraembryonic membranes, are positioned inside the body stalk as it becomes the umbilical cord. **d.** At 35+ days, all membranes are present, and the umbilical cord takes blood vessels between the embryo and the chorion (placenta).

amniotic cavity
embryonic disk
yolk sac
chorionic villi
chorion

a. 18 days

body stalk
allantois
amniotic cavity
embryo
yolk sac
chorionic villi

b. 21 days

placenta
umbilical cord
amnion

Figure 14.12 Fetus and amnion.
Photograph of a human fetus at 8 weeks. The scale bar is 3 cm.

amnion
allantois
yolk sac
chorionic villi
chorion
amniotic cavity

c. 25 days

amniotic cavity
chorion
digestive tract
chorionic villi
amnion
umbilical cord

d. 35+ days

14.4 Fetal Development

During fetal development (last seven months), the skeleton becomes ossified (bony), reproductive organs form, arms and legs develop fully, and the fetus enlarges in size and gains weight.

Three- to four-month-old fetus

Seven- to eight-month-old fetus

Observation: Fetal Development

1. Using Table 14.2 as a guide, examine models of fetal development.
2. In Table 14.2, note the following.
   a. **External genitals:** About the third month, it is possible to tell male from female if an ultrasound is done.
   b. **Quickening:** Fetal movement is felt during the fourth or fifth months.
   c. **Vernix caseosa:** Beginning with the fifth month, the skin is covered with a cheesy coating called vernix caseosa.
   d. **Lanugo:** During the sixth and seventh months, the body is covered with fine, downy hair termed lanugo. – Fetal baby hair →sheds after birth
3. All organ systems become fully functional during fetal development.

Table 14.2	Fetal Development	
Month	**Events for Mother**	**Events for Baby**
Third month	Uterus is the size of a grapefruit.	Possible to distinguish sex. Fingernails appear.
Fourth month	Fetal movement is felt by those who have been previously pregnant. Heartbeat is heard by stethoscope.	Bony skeleton visible. Hair begins to appear. 150 mm (6 in.), 170 g (6 oz.).
Fifth month	Fetal movement is felt by those who have not been previously pregnant. Uterus reaches up to level of umbilicus and pregnancy is obvious.	Protective cheesy coating, called vernix caseosa, begins to be deposited. Heartbeat can be heard.
Sixth month	Doctor can tell where baby's head, back, and limbs are. Breasts have enlarged, nipples and areolae are darkly pigmented, and colostrum is produced.	Body is covered with fine hair called lanugo. Skin is wrinkled and reddish.
Seventh month	Uterus reaches halfway between umbilicus and rib cage.	Testes descend into scrotum. Eyes are open. 300 mm (12 in.), 1,350 g (3 lb).
Eighth month	Weight gain is averaging about a pound a week. Difficulty in standing and walking because center of gravity is thrown forward.	Body hair begins to disappear. Subcutaneous fat begins to be deposited.
Ninth month	Uterus is up to rib cage, causing shortness of breath and heartburn. Sleeping becomes difficult.	Ready for birth. 530 mm (20½ in.), 3,400 g (7½ lb).

Cord Around Baby's Neck

Babies are sometimes born with the umbilical cord wrapped around the neck. There isn't much room in the uterus, and this is a common place for the cord to be located. The umbilical cord around the neck is not as dangerous as it sounds because the baby gets its oxygen by way of the placenta and the umbilical blood vessels and not the lungs. If the umbilical chord becomes overly stretched, oxygen may not be delivered as it should. This causes the fetal heart to slow down. First, the mother can change her position and/or be given oxygen but if the baby's heart slows to below 100 beats for any length of time, then a cesarean delivery may be in order.

Laboratory Review 14

<u>Reproductive</u>   **1.** Where are the testes located in human males?

<u>Conducts sperm</u>   **2.** What is the function of the vas deferens?

<u>   </u>   **3.** What is the function of the prostate gland?

<u>Reproductive</u>   **4.** Where are the ovaries located?

<u>Develop fetus from</u>   **5.** What is the function of the uterus?
embryo

<u>Epididymis</u>   **6.** Where are sperm produced in the testes?

<u>   </u>   **7.** What structure in the ovary contains the developing oocyte (egg)?

<u>   </u>   **8.** Name the stage at which the embryo is a ball of cells.

<u>   </u>   **9.** Which extraembryonic membrane participates in the formation of the placenta?

<u>Cleavage</u>   **10.** What process directly follows the zygote stage?

<u>Starts as cartilage</u> **11.** What overall change in the skeleton occurs during fetal development?
Develops to human skeleton

<u>NO</u>   **12.** Do the lungs function during fetal development?

<u>provides</u>   **13.** What is the function of the placenta?
embryo w/ nutrients / O_2

Thought Questions

14. A vasectomy is a procedure in which both of the vas deferens are severed. Why would such a procedure cause sterility?

15. List the three stages of embryonic development (p. 188, top), and explain.

15

Mitosis and Meiosis

Learning Outcomes

15.1 The Cell Cycle
- Name and describe the stages of the cell cycle.
- Identify the phases of mitosis in models and microscope slides. Explain how the chromosome number stays constant.
- Describe cytokinesis.

Prelab Question: During what stage of the cell cycle does chromosome duplication occur, and why is this critical to mitosis?

15.2 Meiosis
- Name and describe the phases of meiosis I and meiosis II with attention to the movement of chromosomes.
- Explain how the chromosome number is reduced.

Prelab Question: What is synapsis, and why is it critical to meiosis?

15.3 Mitosis Versus Meiosis
- Compare the effects of mitosis to meiosis.
- Contrast the behavior of chromosomes during mitosis with the behavior of chromosomes during meiosis I and II.

Prelab Question: How are mitosis and meiosis II similar? How are they different?

15.4 Gametogenesis
- Contrast spermatogenesis with oogenesis using diagrams and models.

Prelab Question: What differences between spermatogenesis and oogenesis help explain why males produce so many more sperm than females produce eggs?

Application for Daily Living: **Mitosis and Cancer**

Introduction

Dividing cells experience nuclear division, cytoplasmic division, and a period between divisions called interphase. During **interphase,** the nucleus appears normal, and the cell is performing its usual cellular functions. Also, the cell is increasing all of its components, including such organelles as the mitochondria, ribosomes, and centrioles. DNA replication (making an exact copy of the DNA) occurs toward the end of interphase. Thereafter, the chromosomes, which contain DNA, are duplicated and contain two chromatids held together at a **centromere.** These chromatids are called **sister chromatids.**

When the nucleus divides during **mitosis,** the daughter nuclei receive the same number of chromosomes and genetic material as the parent cell. During cytokinesis, the cytoplasm divides and two daughter cells are produced. Mitosis in humans permits growth and repair of tissues. During sexual reproduction, another form of division called **meiosis** occurs. Meiosis is a part of **gametogenesis,** the production of gametes (sex cells called sperm in males and eggs in females). As a result of meiosis, the daughter cells have half the number of chromosomes as the parent cell. Also, the chromatids sometimes exchange genetic material during crossing-over; therefore, the daughter cells do not have the same number of chromosomes and are not genetically identical to the parent cell, following meiosis.

15.1 The Cell Cycle

As stated in the introduction, the period between cell divisions is known as interphase. Early investigators noted little visible activity between cell divisions, so they dismissed this period as a resting state. But when they discovered that DNA replication occurs (and the chromosomes become duplicated) during interphase, the **cell cycle** concept was proposed. Investigators have also discovered that cytoplasmic organelle duplication occurs during G_1 and synthesis of the proteins involved in regulating cell division occurs during G_2. Thus, the cell cycle can be broken down into four stages (Fig. 15.1). State the event of each stage on the line provided:

G_1 _____

S _____

G_2 _____

M _____

Why is the entire process called the "cell cycle"? _____

_____

Figure 15.1 The cell cycle.
Immature cells repeatedly go through a cycle that consists of four stages: G1, S (for synthesis), G2, and M (for mitosis). Eventually, some daughter cells "break out" of the cell cycle and become specialized cells.

Duration of the Cell Cycle

If mature human cells are speedily dividing, the length of the cell cycle is about 22–24 hours. Embryonic cells go through the cell cycle much more rapidly by omitting G_1 and G_2, so that the cell cycle includes only DNA replication and mitosis. Mature cells usually take this amount of time per phase: G_1 = 10 hours; S = 5–6 hours; G_2 = 3–4 hours; mitosis and cytokinesis = 2 hours.

Mitosis

Mitosis is nuclear division that results in two new nuclei, each having the same number of chromosomes as the original nucleus. The **parent cell** is the cell that divides, and the resulting cells are called **daughter cells.** If a parent cell has 46 chromosomes, how many chromosomes does each daughter cell have following mitosis?

_____

When cell division is about to begin, chromatin starts to condense and compact to form visible, rodlike **sister chromatids** held together at the centromere (Fig. 15.2a). *Label the sister chromatids and the centromere in Figure 15.2b.* This illustration represents a **duplicated chromosome** as it would appear just before nuclear division occurs.

Spindle

Table 15.1 lists the structures that play a role during mitosis. The **spindle** is a structure that appears and brings about an orderly distribution of daughter chromosomes to the daughter cell nuclei. A spindle has fibers that stretch between two poles (ends) (Fig. 15.3). **Spindle fibers** are bundles of microtubules, protein cylinders found in the cytoplasm that can assemble and disassemble. The **centrosome,** which is the main microtubule-organizing center of the cell, divides before mitosis so that each pole of the spindle has a pair of centrosomes. Human cells contain two barrel-shaped organelles called **centrioles** in each centrosome and **asters,** arrays of short microtubules radiating from the poles. Thus, centrioles mark the location of the spindle poles.

Figure 15.2   Duplicated chromosomes.
DNA replication results in a duplicated chromosome that consists of two sister chromatids held together at a centromere. **a.** Scanning electron micrograph of a duplicated chromosome. **b.** Drawing of a duplicated chromosome. During mitosis, each chromatid becomes a daughter chromosome.

1. _____

2. _____

one chromatid

a.

b.

Table 15.1	Structures Associated with Mitosis
Structure	**Description**
Nucleus	A large organelle containing the chromosomes and acting as a control center for the cells
Chromosome	Rod-shaped body in the nucleus seen during mitosis and meiosis that contains DNA and therefore the hereditary units, or genes
Nucleolus	An organelle found inside the nucleus; composed largely of RNA for ribosome formation
Spindle	Microtubule structure that brings about chromosome movement during nuclear division
Chromatids	The two identical parts of a chromosome following DNA replication; they become daughter chromosomes
Centromere	A constriction where duplicates (sister chromatids) of a chromosome are held together
Centrosome	The central microtubule-organizing center of cells; consists of granular material; in human cells, contains two centrioles
Centriole	A short, cylindrical organelle in human cells that contains microtubules. They mark the location of the spindle poles.
Aster	Short, radiating fibers produced by the centrioles; important during mitosis and meiosis

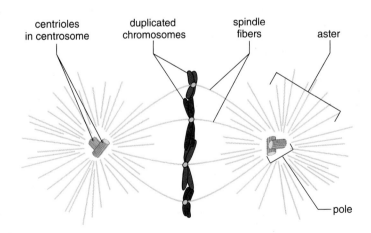

centrioles in centrosome   duplicated chromosomes   spindle fibers   aster

pole

Figure 15.3   The spindle.
Spindle fibers (bundles of microtubules) help distribute the daughter chromosomes during mitosis.

Figure 15.4   Phases of mitosis in human cells.

This illustration extends across two pages. The colors signify that the chromosomes were inherited from different parents.

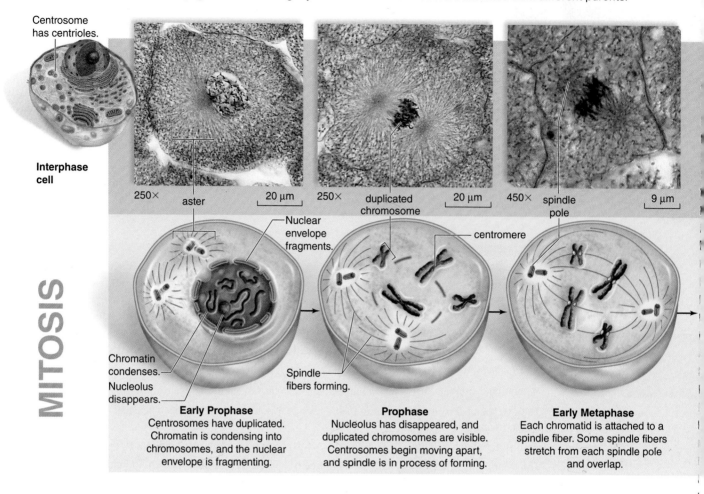

Centrosome has centrioles.

Interphase cell

250×   aster   20 μm   250×   duplicated chromosome   20 μm   450×   spindle pole   9 μm

MITOSIS

Nuclear envelope fragments.

centromere

Chromatin condenses.
Nucleolus disappears.

Spindle fibers forming.

Early Prophase
Centrosomes have duplicated.
Chromatin is condensing into chromosomes, and the nuclear envelope is fragmenting.

Prophase
Nucleolus has disappeared, and duplicated chromosomes are visible. Centrosomes begin moving apart, and spindle is in process of forming.

Early Metaphase
Each chromatid is attached to a spindle fiber. Some spindle fibers stretch from each spindle pole and overlap.

Observation: Mitosis

cell wall won't form on animal cells

Mitosis Models

1. Using the descriptions given in Figure 15.4 as a guide, identify the phases of human cell mitosis in models of human cell mitosis.
2. Each species has its own chromosome number. Counting the number of centromeres tells you the number of chromosomes in the models. What is the number of chromosomes observed in each nucleus of the cells? _____

Whitefish Blastula Slide

The blastula is an early embryonic stage in the development of humans. The **blastomeres** (blastula cells) that are in the top row of Figure 15.4 are in the illustrated phases of mitosis.

1. Examine a prepared slide of whitefish blastula cells undergoing mitotic cell division.
2. Try to find a cell in each phase of mitosis. Have a partner or your instructor check your identification.
3. Determine if you find more cells in prophase than the other phases. If so, what does that tell you about the length of time a cell stays in prophase? _____

_____

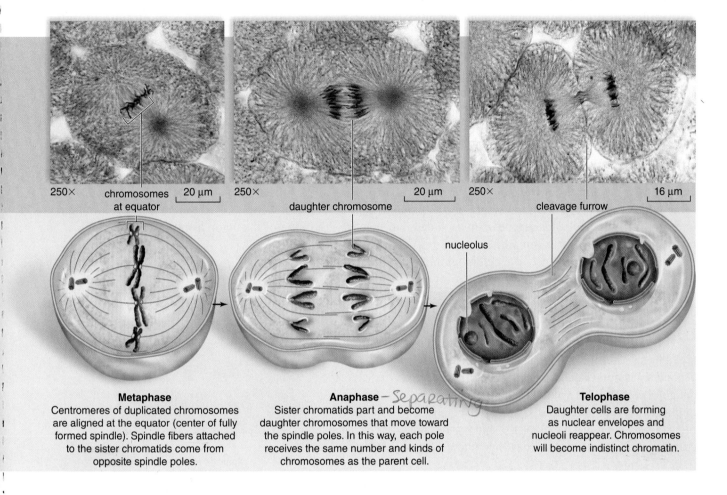

250× chromosomes 20 µm 250×   20 µm 250×   16 µm
        at equator   daughter chromosome   cleavage furrow

nucleolus

Metaphase
Centromeres of duplicated chromosomes are aligned at the equator (center of fully formed spindle). Spindle fibers attached to the sister chromatids come from opposite spindle poles.

Anaphase — Separating
Sister chromatids part and become daughter chromosomes that move toward the spindle poles. In this way, each pole receives the same number and kinds of chromosomes as the parent cell.

Telophase
Daughter cells are forming as nuclear envelopes and nucleoli reappear. Chromosomes will become indistinct chromatin.

Mitosis Phases = NUCLEAR DIVISION —

The phases of mitosis are **prophase, metaphase, anaphase,** and **telophase**—in that order (Fig. 15.4). Early stages of prophase and metaphase are also shown in this figure.

Prophase

During early prophase, the chromosomes continue to condense, the **nucleolus** disappears, and the nuclear envelope fragments. Still during prophase, the chromosomes have no apparent orientation within the cell. The already duplicated chromosomes are composed of two sister chromatids held together at a centromere. Counting the number of centromeres in diagrammatic drawings gives the number of chromosomes for the cell. What is the

chromosome number for the cells in Figure 15.4? _____

The spindle begins to assemble as the centrosomes, each containing two centrioles, migrate to the poles.

Contrast the appearance of a cell in prophase with one in interphase. (*Hint:* See Fig. 15.1 and compare the

nuclei and their contents.) _____

_____

Metaphase

The sister chromatids are now attached to the spindle, and the chromosomes are aligned at the equator of the spindle. The mitotic spindle occupies the region formerly occupied by the nucleus. Short microtubules radiate out in a starlike aster from the pair of centrioles located in each centrosome. The spindle consists of poles, asters, and fibers, bundles of parallel microtubules.

Contrast the appearance of a cell in metaphase with one in prophase. _____

_____

Anaphase

At the start of anaphase, the centromeres split, and the sister chromatids of each chromosome separate, giving rise to two daughter chromosomes. The daughter chromosomes begin to move toward opposite poles of the spindle. Each pole receives the diploid number of daughter chromosomes.

     Contrast the appearance of a cell in anaphase with one in metaphase. _____

_____

Telophase

New nuclear envelopes form around the daughter chromosomes at the poles. Each daughter nucleus contains the same number and types of chromosomes as the parental cell. The chromosomes become more diffuse chromatin once again, and a nucleolus reappears in each daughter nucleus. Division of the cytoplasm by formation of a **cleavage furrow** is nearly complete.

     Contrast the appearance of a cell in telophase with one in anaphase. _____

_____

Cytokinesis - Cytoplasmic Division

Cytokinesis, division of the cytoplasm, usually accompanies mitosis. During cytokinesis, each daughter cell receives a share of the organelles that duplicated during interphase. Cytokinesis begins in anaphase, continues in telophase, and reaches completion by the start of the next interphase.

Cytokinesis in Human Cells

In human cells, a cleavage furrow, an indentation of the membrane between the daughter nuclei, begins as anaphase draws to a close. The cleavage furrow deepens as a band of actin filaments called the contractile ring slowly constricts the cell, forming two daughter cells (Fig. 15.5).

     Were any of the cells of the whitefish

blastula slide undergoing cytokinesis? _____

How do you know? _____

_____

Summary of Mitotic Cell Division

1. The nuclei in the daughter cells have the

   _____ number of chromosomes as the parent cell had.

2. Mitosis is cell division in which the

   chromosome number _____

   _____

4,000×

cleavage furrow

contractile ring

Figure 15.5 Cytokinesis in human cells.
A single cell becomes two cells by a furrowing process. A contractile ring composed of actin filaments gradually gets smaller, and the cleavage furrow pinches the cell into two cells.

15.2 Meiosis

Meiosis is a form of nuclear division in which the chromosome number is reduced by half. The nucleus of the parent cell has the **diploid (2n) number** of chromosomes; the daughter nuclei, after meiosis is complete, have the **haploid (n) number** of chromosomes. In sexually reproducing species, meiosis must occur or the chromosome number would double with each generation.

A diploid cell nucleus contains **homologues,** also called homologous chromosomes. The homologues of each pair look alike and carry the genes for the same traits. Before meiosis begins, the chromosomes are already double stranded—they contain sister chromatids. Meiosis requires two divisions, called **meiosis I** and **meiosis II** (Fig. 15.6).

Meiosis I

During prophase of meiosis I, the spindle appears and the nuclear envelope and nucleolus disappear. Homologues line up next to one another during a process called synapsis. During **crossing-over,** the nonsister chromatids of a homologue pair exchange genetic material. At metaphase I, the homologue pairs line up at the equator of the spindle. During anaphase I, homologues separate and the chromosomes (still having two chromatids) move to each pole. In telophase I, the nuclear envelope and the nucleolus reappear as the spindle disappears. Each new nucleus contains one homologue from each pair of homologues.

Prophase I

Use Figure 15.6 to learn the major events of each phase of meiosis I. In prophase I the chromosomes are in a fragmented nuclear envelope and the homologues undergo synapsis and nonsister chromatids undergo crossing-over. (The red-short and blue-short are homologues; red-long and blue-long are also homologues.)

Metaphase I

The homologues are at the equator, prepared to move apart toward the poles. Each homologue pair acts independently and either homologue can be facing either pole. Notice that crossing-over is represented by an exchange of color between nonsister chromatids. Why do nonsister chromatids participate in crossing-over but sister

chromatids do not? _____

_____

(The difference in color between nonsister chromatids represents different genetic material.)

Anaphase I

The members of each homologue pair separate, and they move toward opposite poles. Now the nuclei will be haploid: they no longer have homologue pairs.

Telophase I

1. During telophase I, the chromosomes are at the poles. Disregard any crossing-over and state what combinations of chromosomes are at the poles? Fill in the following blanks with the words *red-long, red-short, blue-long,* and *blue-short:*

   Pole A: _____ and _____

   Pole B: _____ and _____

2. What other combinations would have been possible? (*Hint:* Alternate the colors at metaphase I.)

   Pole A: _____ and _____

   Pole B: _____ and _____

   All possible combinations of chromosomes can occur in daughter nuclei following meiosis I. Why would you not find two short chromosomes nor two long chromosomes at the poles of the spindle following

   telophase I? _____

Figure 15.6 Meiosis I and II in human cell drawings.

This illustration extends across two pages. Meiosis I begins on this page and continues on the next page; meiosis II begins on this page and continues on the next page. Note that only by coloring the homologues differently (red and blue) is it possible to show that the daughter cells following meiosis I vary genetically and to show the increased variation caused by crossing-over.

Interphase cell

Centrosome has centrioles.

2n = 4

Prophase I
Chromosomes have duplicated. Homologue pair during synapsis and crossing-over occurs.

Metaphase I
Homologue pairs align independently at the equator.

kinetochore

Anaphase I
Homologues separate and move toward the poles.

MEIOSIS I

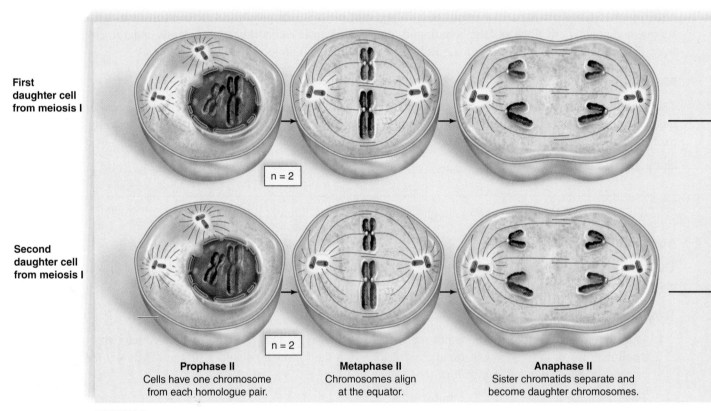

First daughter cell from meiosis I

Second daughter cell from meiosis I

n = 2

n = 2

Prophase II
Cells have one chromosome from each homologue pair.

Metaphase II
Chromosomes align at the equator.

Anaphase II
Sister chromatids separate and become daughter chromosomes.

MEIOSIS II

Telophase I
Daughter cells have one homologus from each homologue pair.

Interkinesis
Chromosomes still consist of two chromatids.

First daughter cell

Second daughter cell

n = 2

n = 2

MEIOSIS I cont'd

n = 2

Telophase II
Spindle disappears, nuclei form, and cytokinesis takes place.

Daughter cells
Meiosis results in four haploid daughter cells.

n = 2

MEIOSIS II cont'd

Events of Meiosis I

Show the events of meiosis I by adding two pairs of simplified homologues (chromatids can be straight lines) to Figure 15.7. Remember that the members of a homologue pair are colored differently (red/blue) and are recognized by length. Assume that nonsister chromatids have already experienced crossing-over when you begin.

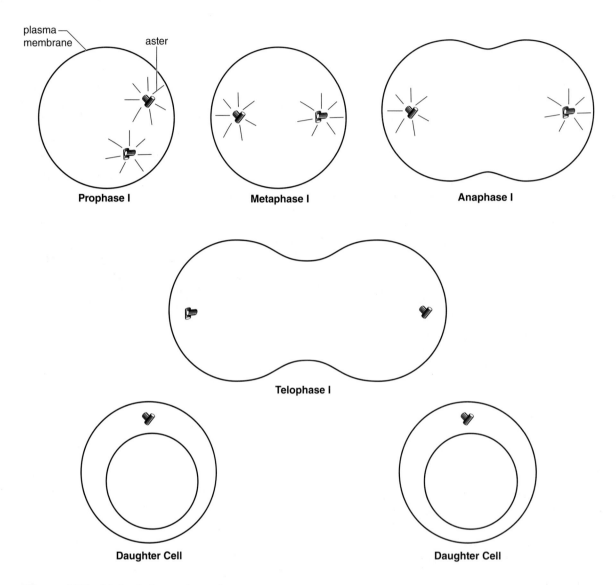

Figure 15.7   Meiosis I exercise art.

Conclusions: Meiosis I

- Do the chromosomes inherited from the mother (e.g., red) or father (e.g., blue) have to remain together following meiosis I? _____ Does the same genetic material have to remain together? _____

- Name two ways that meiosis contributes to genetic recombination:

  a. _____

  b. _____

Interkinesis

Interkinesis is the period between meiosis I and meiosis II. In some species, daughter cells do not form and meiosis II follows right after meiosis I. Does DNA replication occur during interkinesis prior to meiosis II? _____ Explain why not. _____

_____

Meiosis II

During prophase of meiosis II, a spindle appears. Each chromosome individually attaches to the spindle. During metaphase II, the chromosomes are lined up at the equator. During anaphase II, the centromeres divide and the chromatids separate, becoming daughter chromosomes that move toward the poles. In telophase II, the spindle disappears as the nuclear envelope reappears. Meiosis II is exactly like mitosis (see Fig. 15.4) except that the nuclei of the parent cell and the daughter cells are haploid.

   If so directed by your instructor, use the space below to create a series of drawings showing the stages of meiosis II. Otherwise, simply answer these questions. In Figure 15.6, how many chromosomes are in the nucleus during each phase of meiosis II, whether prophase II, metaphase II, telophase II, or the daughter cells? _____ Explain how this is possible. _____

_____

_____

Summary of Meiotic Cell Division

1. The parent cell has the diploid (2n) number of chromosomes, and the daughter cells following meiosis have the _____(n) number of chromosomes.

2. Meiosis is cell division in which the chromosome number _____.

3. If a parent cell has 16 chromosomes, the daughter cells will have how many chromosomes following meiosis? _____

4. Whereas meiosis reduces the chromosome number, **fertilization** restores the chromosome number. A zygote contains the same number of chromosomes as the parent, but are these exactly the same chromosomes? _____

5. What is another way that sexual reproduction results in genetic recombination? _____

   _____

   _____

15.3 Mitosis Versus Meiosis

In this section we will utilize Figure 15.8 to (1) compare the process of mitosis to the process of meiosis in general and then (2) to specifically compare mitosis to meiosis I and finally (3) to specifically compare mitosis to meiosis II.

General Differences

To fill in Table 15.2, look at the completion of mitosis and meiosis II on page 211 to see how many divisions were required for mitosis versus meiosis and the chromosome number (2n/n) of the daughter cells and the number of daughter cells for each process. Now fill in Table 15.2

Table 15.2 Differences Between Mitosis and Meiosis		
	Mitosis	**Meiosis**
1. Number of divisions		
2. Chromosome number in daughter cells		
3. Number of daughter cells		

Figure 15.8   Meiosis I and meiosis II compared to mitosis.
Compare metaphase I of meiosis I to metaphase of mitosis. Only in metaphase I are the homologues paired at the equator. Members of each homologue pair separate during anaphase I, and therefore the daughter cells are haploid. The exchange of color between nonsister chromatids represents the crossing-over that occurred during meiosis I.

MEIOSIS I

haploid

Prophase I	Metaphase I	Anaphase I
2n = 4		
Synapsis and crossing-over occur.	Homologue pairs align independently at the equator.	Homologues separate and move toward the poles.

MITOSIS

Diploid cells

Prophase	Metaphase	Anaphase
2n = 4		
	Chromosomes align at the equator.	Sister chromatids separate and become daughter chromosomes.

Specific Differences

To compare the specifics of mitosis to meiosis I, notice that phrases in the first column of Table 15.3 are correct for mitosis. If the phrase also applies to meiosis I, just repeat it in the second column. If it doesn't apply, correct it for meiosis I. To get you started, the correct phrase has been supplied for prophase I.

Table 15.3 Mitosis Compared with Meiosis I	
Mitosis	**Meiosis I**
Prophase: no pairing of chromosomes	Prophase I: _____ Pairing of homologues _____
Metaphase: duplicated chromosomes at equator	Metaphase I: _____
Anaphase: sister chromatids separate.	Anaphase I: _____
Telophase: chromosomes have one chromatid.	Telophase I: _____

To compare mitosis to meiosis II, notice that the phrases in the first column of Table 15.4 are correct for mitosis. If the phrase also applies to meiosis II (see one daughter cell at the bottom of Fig. 15.6), just repeat it in the second column. To get you started, the correct phrase has been supplied for prophase II.

Table 15.4 Mitosis Compared with Meiosis II

Mitosis	Meiosis II
Prophase: no pairing of chromosomes	Prophase II: _____ No pairing of chromosomes _____
Metaphase: duplicated chromosomes at equator	Metaphase II: _____
Anaphase: sister chromatids separate.	Anaphase II: _____
Telophase: two diploid daughter cells	Telophase II: _____

Change in Chromosome Structure

Anomalies of chromosome structure can occur when cells divide, particularly if cells have been subject to environmental influences such as radiation or drug intake. Some of the more common structural anomalies that can be seen in a karyotype are

Deletion: The chromosome is shorter than usual because some portion is missing.
Duplication: The chromosome is longer than usual because some portion is present twice over.
Inversion: The chromosome is normal in length but some portion runs in the opposite direction.
Translocation: Two chromosomes have switched portions and each switched portion is on the wrong chromosome.

Anomalies of chromosome structure can also result in recognized syndromes. A **syndrome** is characterized by a group of symptoms and conditions that always occur together (see page 225).

15.4 Gametogenesis

Gametogenesis is the formation of **gametes** (sex cells), the sperm and egg. **Fertilization** occurs when the nucleus of a sperm fuses with the nucleus of an egg.

Gametogenesis

Gametogenesis occurs in the testes of males, where **spermatogenesis** produces sperm. Gametogenesis occurs in the ovaries of females, where **oogenesis** produces oocytes (eggs).

Recall that a diploid (2n) nucleus contains the full number of chromosomes, and a haploid (n) nucleus contains half as many. Gametogenesis involves meiosis, the process that reduces the chromosome number from 2n to n. In sexually reproducing species, if meiosis did not occur, the chromosome number would double with each generation. Because meiosis consists of two divisions—the first meiotic division (meiosis I) and the second meiotic division (meiosis II)—you expect four haploid cells at the end of the process. Indeed, there are four sperm as a result of spermatogenesis (Fig. 15.9). However, in females, meiosis I results in a secondary oocyte and one polar body.

A **polar body** is a nonfunctioning cell that will disintegrate. A secondary oocyte does not undergo meiosis II unless fertilization (fusion of egg and sperm) occurs. At the completion of oogenesis, there is a single egg. The polar bodies die (Fig. 15.9). The egg has most of the cytoplasm.

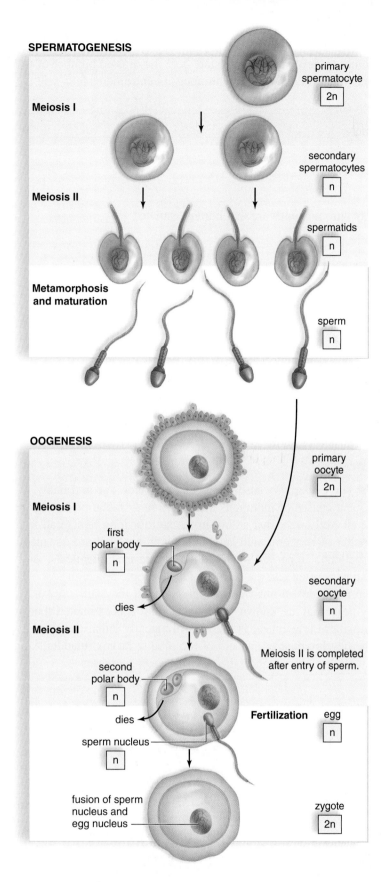

SPERMATOGENESIS

primary spermatocyte
2n

Meiosis I

secondary spermatocytes
n

Meiosis II

spermatids
n

Metamorphosis and maturation

sperm
n

OOGENESIS

primary oocyte
2n

Meiosis I

first polar body
n

dies

secondary oocyte
n

Meiosis II

Meiosis II is completed after entry of sperm.

second polar body
n

dies

Fertilization   egg
n

sperm nucleus
n

fusion of sperm nucleus and egg nucleus

zygote
2n

Figure 15.9   Spermatogenesis and oogenesis.
Spermatogenesis produces four viable sperm, whereas oogenesis produces one egg and two polar bodies. In humans, both sperm and egg have 23 chromosomes each; therefore, following fertilization, the zygote has 46 chromosomes.

Observation: Gametogenesis

Examine any available gametogenesis models and state here the diploid number of the parent cell and the haploid number of a gamete. Remember that counting the number of centromeres tells you the number of chromosomes.

Diploid number of parent cell: _____

Haploid number of a gamete: _____

Summary of Gametogenesis

1. What is gametogenesis? _____

   In general, how many chromosomes are in a human gamete? _____

2. What is spermatogenesis? _____

   How many chromosomes does a human sperm have? _____

3. What is oogenesis? _____

   How many chromosomes does a human egg have? _____

4. Following fertilization, how many chromosomes does the zygote, the first cell of the new

   individual, have? _____

Application for Daily Living

Mitosis and Cancer

Mutations can upset the cell cycle and cause body cells to divide uncontrollably. Such mutations often lead to cancer. Two of the frequent causes of mutations are exposure to radiation or organic chemicals. Radiation is in sunlight, and we are all well aware that sitting in the sun for long hours can lead to skin cancer. Frequent X-rays can also be a matter of concern, and we should avoid any that are not medically necessary. Mutagenic organic chemicals can be found in certain food additives, industrial chemicals, and pesticides. That's why lawns sprayed with pesticides often carry a warning label. People sometimes show extreme concern about industrial chemicals and pesticides in our water supply, and yet they still smoke. Tobacco smoke contains a number of organic chemicals that are known carcinogens, and it is estimated that one-third of cancer deaths can be attributed to smoking. Lung cancer is the most frequently lethal cancer in the United States; smoking is also implicated in the development of cancers of the mouth, larynx, bladder, kidney, and pancreas. When smoking is combined with drinking alcohol, the risk of cancer increases.

Interphase - S

1. During what stage of the cell cycle does DNA replication occur?

Anaphase

2. Name the phase of mitosis during which separation of sister chromatids occurs.

Cytokinesis

3. By what process does the cytoplasm of a human cell separate?

Prophase

4. Name the phase of mitosis in which duplicated chromosomes first appear. *For mitosis!*

Reproductive

5. Where in humans would you expect to find meiosis taking place? *The DNA/chromosom is duplicated in "S" stage.*

13

6. If there are 13 pairs of homologues at the start of spermatogenesis, how many chromosomes are there in a sperm?

oogenesis

7. What term refers to the production of an egg? one cell

oogenesis

8. During which type of gametogenesis would you see polar bodies?

homologues

9. What do you call chromosomes that have the same length and carry genes for the same traits? homologues

Anaphase

10. If homologues are separating, what phase is this?

12

11. If the parent cell has 24 chromosomes, how many does each daughter cell have at the completion of meiosis II? 2n 2(12)=24

meiosis

12. Name the type of cell division during which homologues pair.

mitosis

13. Name the type of cell division described by 2n → 2n.

yes

14. Does metaphase of mitosis, meiosis I, or meiosis II have the haploid number of duplicated chromosomes at the equator of the spindle?

Thought Questions

15. Meiosis functions to reduce chromosome number. When, during the human life cycle, is the diploid number of chromosomes restored?

16. How does the alignment of chromosomes differ between metaphase of mitosis and metaphase of meiosis I?

17. A student is simulating meiosis I with homologues that are red-long and yellow-long. Describe the appearance of two nonsister chromatids following crossing-over.

McGraw-Hill Access Science Website

An online encyclopedia of science and technology that provides information, including videos, that can enhance the laboratory experience.

www.accessscience.com

connect | BIOLOGY

Find even more resources to learn the chapter concepts including animations, tutorial videos, and interactive practice questions.

LEARNSMART LABS®

The following LearnSmart Lab contains exercises that are related to the content of this chapter:

• Mitosis & Meiosis

16

Patterns of Genetic Inheritance

Learning Outcomes

16.1 Determining the Genotype
- Determine the genotype by observation of the person and their relatives.

Prelab Question: Both Nancy and Jim are homozygous for freckles, a dominant trait. Using letters, what is their genotype?

16.2 Determining Inheritance
- Do genetics problems involving autosomal dominant, autosomal recessive, and X-linked recessive alleles.
- Do genetic problems involving multiple allele inheritance and use blood type to help determine paternity.

Prelab Question: What are the chances that Nancy and Jim will have children with freckles?

16.3 Genetic Counseling
- Determine whether a karyotype contains a chromosome anomaly.
- Determine whether a pedigree represents a pattern of autosomal dominant, autosomal recessive, or X-linked recessive inheritance.
- Construct a pedigree to determine the chances of inheriting a particular phenotype when provided with generational information.

Prelab Question: In the pedigree of Nancy's family and Jim's family, would you expect to find few or many people with freckles? Explain.

Application for Daily Living: **Choosing the Gender of Your Child**

Introduction

Particulate genetics, which views the genes as particles located on the chromosomes, is no longer supported by molecular genetics. Still, its principles can be used to understand patterns of inheritance. First, a gene has two alternate forms called **alleles** for any trait such as hair line, finger length, and so on. One possible allele designated by a capital letter is **dominant** over the **recessive** allele, designated by a lowercase letter. An individual can be **homozygous dominant** (two dominant alleles, *EE*), **homozygous recessive** (two recessive alleles, *ee*) or **heterozygous** (one dominant and one recessive allele, *Ee*). **Genotype** refers to an individual's alleles, and **phenotype** refers to an individual's appearance (Fig. 16.1). Homozygous dominant and also heterozygous individuals show the dominant phenotype; homozygous recessive individuals show the recessive phenotype.

Figure 16.1 Genotype versus phenotype.
Unattached earlobes (*E*) are dominant over attached earlobes (*e*). **a.** Homozygous dominant individuals have unattached earlobes. **b.** Homozygous recessive individuals have attached earlobes. **c.** Heterozygous individuals have unattached earlobes.

a. Unattached earlobe

b. Attached earlobe

c. Unattached earlobe

16.1 Determining the Genotype

Autosomal traits are determined by alleles on the autosomal chromosomes, all chromosomes except the sex chromosomes.

Autosomal Dominant and Recessive Traits

Figure 16.2 shows a few human traits.

1. What is the homozygous dominant genotype for type of hairline? _____ What is the

    phenotype? _____

2. What is the homozygous recessive genotype for finger length? What is the phenotype?

    _____

Figure 16.2 Commonly inherited traits in human beings.
The alleles indicate which traits are dominant and which are recessive.

a. Widow's peak: *WW* or *Ww*   **b.** Straight hairline: *ww*   **e.** Short fingers: *SS* or *Ss*   **f.** Long fingers: *ss*

c. Unattached earlobes: *EE* or *Ee*   **d.** Attached earlobes: *ee*   **g.** Freckles: *FF* or *Ff*   **h.** No freckles: *ff*

3. Why does the heterozygous individual *Ff* have freckles? _____

These genetic problems use the alleles from Figure 16.2 and Table 16.1.

4. Victoria and the members of her immediate family have attached earlobes. What is Victoria's genotype?

    _____ Her maternal grandfather has unattached earlobes. Deduce the genotype of her maternal

    grandfather. _____ Explain. _____

    _____

5. Lucas does not have a bent little finger, but his parents do. Deduce the genotype of his parents. _____

    Of Lucas. _____ Explain. _____

6. Axel is adopted. He has hair on the back of his hand. Could both of his parents have had hair on the

    back of the hand? _____ Could both of his parents have had no hair on the back of the hand?

    _____ Explain. _____

1. For this Experimental Procedure, you will need a lab partner to help you determine your phenotype for the traits listed in the first column of Table 16.1.

2. Determine your probable genotype. If you have the recessive phenotype, you know your genotype. If you have the dominant phenotype, you may be able to decide whether you are homozygous dominant or heterozygous by recalling the phenotype of your parents, siblings, or children. *Circle your probable genotype in the second column of Table 16.1.*

3. Your instructor will tally the class's phenotypes for each trait so that you can complete the third column of Table 16.1.

4. Complete Table 16.1 by calculating the percentage of the class with each trait. Are dominant phenotypes always the most common in a population? _____ Explain. _____

_____

Table 16.1   Autosomal Human Traits

Trait: d = Dominant r = Recessive	Possible Genotypes	Number in Class	Percentage of Class with Trait
Hairline:			
Widow's peak (d)	*WW* or *Ww*	_____	_____
Straight hairline (r)	*ww*	_____	_____
Earlobes:			
Unattached (d)	*UU* or *Uu*	_____	_____
Attached (r)	*uu*	_____	_____
Skin pigmentation:			
Freckles (d)	*FF* or *Ff*	_____	_____
No freckles (r)	*ff*	_____	_____
Hair on back of hand:			
Present (d)	*HH* or *Hh*	_____	_____
Absent (r)	*hh*	_____	_____
Thumb hyperextension—"hitchhiker's thumb":			
Last segment cannot be bent backward (d).	*TT* or *Tt*	_____	_____
Last segment can be bent back to 60° (r).	*tt*	_____	_____
Bent little finger:			
Little finger bends toward ring finger (d).	*LL* or *Ll*	_____	_____
Straight little finger (r)	*ll*	_____	_____
Interlacing of fingers:			
Left thumb over right (d)	*II* or *Ii*	_____	_____
Right thumb over left (r)	*ii*	_____	_____

16.2 Determining Inheritance

A Punnett square is a means to determine the genetic inheritance of offspring if the genotypes of both parents are known. Figure 16.3 shows how to construct a Punnett square when autosomal alleles are involved. In a **Punnett square**, all possible types of sperm are lined up vertically, and all possible types of eggs are lined up horizontally, or vice versa, so that every possible combination of gametes occurs within the squares.

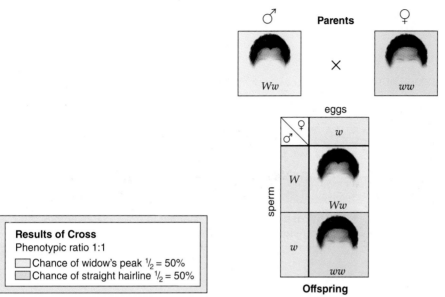

Results of Cross
Phenotypic ratio 1:1
☐ Chance of widow's peak ½ = 50%
☐ Chance of straight hairline ½ = 50%

Figure 16.3   Punnett square.

Calculating the Chances of a Phenotype

Same as in Figure 16.3, learn to calculate the chances that an offspring will have a certain phenotype.

If in a Punnett Square, ¼ of the offspring have the phenotype = _____% chance.

If in a Punnett Square, ¾ of the offspring have the phenotype = _____% chance.

If in a Punnett Square, ½ of the offspring have the phenotype = _____% chance.

In these genetics problems, use letters to fill in the parentheses with the genotype of the parents.

1. **a.** With reference to Figure 16.4*a,* if a genetic disorder is *recessive* and both parents are
heterozygous (_____), what are the chances that an offspring will have the disorder? _____

   b. With reference to Figure 16.4*a,* if a genetic disorder is *dominant* and both parents are
heterozygous (_____), what are the chances that an offspring will have the disorder? _____

Figure 16.4 Two common patterns of autosomal inheritance in humans.

a. Both parents are heterozygous. **b.** One parent is heterozygous and the other is homozygous recessive. The letter *A* stands for any trait that is dominant and the letter *a* stands for any trait that is recessive. Substitute in the correct alleles for the problem you are working on. For example *C* = normal; *c* = cystic fibrosis.

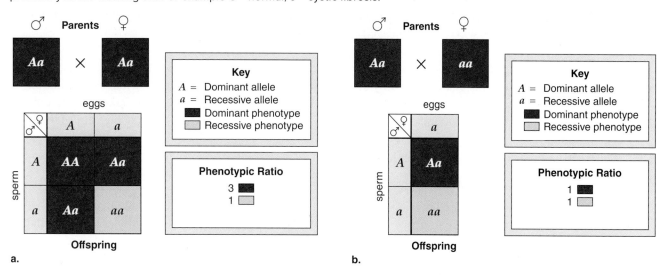

2. **a.** With reference to Figure 16.4*b*, if the parents are heterozygous (_____) by homozygous recessive

   (_____), and the genetic disorder is *recessive*, what are the chances that the offspring will have the

   disorder? _____

   b. With reference to Figure 16.4*b*, if the parents are heterozygous (_____) by homozygous recessive

   (_____), and the genetic disorder is *dominant*, what are the chances that an offspring will have the

   disorder? _____

Autosomal Disorders

1. **Neurofibromatosis (NF),** sometimes called von Recklinghausen disease, is one of the most common genetic disorders. It affects roughly 1 in 3,000 people. It is seen equally in every racial and ethnic group throughout the world. At birth or later, the affected individual may have six or more large tan spots on the skin. Such spots may increase in size and number and become darker. Small benign tumors (lumps) called neurofibromas may occur under the skin or in the muscles. Neurofibromas are made up of nerve cells and other cell types.

   Neurofibromatosis is a dominant disorder. If a heterozygous woman reproduces with a homozygous

   normal man, what are the chances a child will have neurofibromatosis? _____

2. **Cystic fibrosis** is due to abnormal mucus-secreting tissues. At first, the newborn infant may have difficulty regaining the birth weight despite good appetite and vigor. A cough associated with a rapid respiratory rate but no fever indicates lung involvement. Large, frequent, and foul-smelling stools are due to abnormal pancreatic secretions. Whereas children previously died in infancy due to infections, they now often survive because of antibiotic therapy.

   Cystic fibrosis is a recessive disorder. A **carrier** is an individual that appears to be normal but carries

   a recessive allele for a genetic disorder. A man and a woman are both carriers (_____) for cystic

   fibrosis. What are the chances their child will have cystic fibrosis? _____

3. **Huntington disease** does not appear until the 30s or early 40s. There is a progressive deterioration of the individual's nervous system that eventually leads to constant thrashing and writhing movements until insanity precedes death. Studies suggest that Huntington disease is due to a single faulty gene that has multiple effects, in which case there is now hope for a cure.

People with Huntington disease seem to be more fertile than others. It is amazing that more than 1,000 of the cases in the United States in the past century can be traced to one man born in 1831. Huntington disease is a dominant disorder. Sophia is 25 years old and as yet has no signs of Huntington disease. Her mother does have Huntington disease (_____), but her father is free (_____) of the disorder. What are the chances that Sophia will develop Huntington disease?

_____

4. **Phenylketonuria (PKU)** is characterized by severe mental retardation due to an abnormal accumulation of the common amino acid phenylalanine within cells, including neurons. The disorder takes its name from the presence of a breakdown product, phenylketone, in the urine and blood. Newborn babies are routinely tested at the hospital and, if necessary, are placed on a diet low in phenylalanine.

Phenylketonuria (PKU) is a recessive disorder. Mr. and Mrs. Rodriquez appear to be normal, but they have a child with PKU. What are the genotypes of Mr. and Mrs. Rodriquez? _____

5. **Tay–Sachs disease** is caused by the inability to break down a certain type of fat molecule that accumulates around nerve cells until they are destroyed. Afflicted newborns appear normal and healthy at birth, but they do not develop normally. At first, they may learn to sit up and stand, but later they regress and become mentally retarded, blind, and paralyzed. Death usually occurs between ages three and four.

Tay–Sachs is an autosomal recessive disorder. Is it possible for two individuals who do not have Tay–Sachs to have a child with the disorder? Explain. _____

X-Linked Inheritance and Disorders

The sex chromosomes designated X and Y carry genes just like the autosomal chromosomes. Genes particularly on the X chromosome that have nothing to do with gender inheritance are said to be X-linked. **X-linked recessive disorders** are due to recessive genes carried on the X chromosomes. Males are more likely to have an X-linked recessive disorder than females because the Y chromosome is blank for this trait. Does a color-blind male give his son a recessive-bearing X or a Y that is blank for the recessive allele? _____

The possible genotypes and phenotypes for an X-linked color blindness are as follows:

Females
$X^B X^B$ = normal vision
$X^B X^b$ = normal vision (carrier)
$X^b X^b$ = color blindness

Males
$X^B Y$ = normal vision
$X^b Y$ = color blindness

An X-linked recessive disorder in males is always inherited from his mother. Most likely, his mother is heterozygous and therefore does not show the disorder. She is designated a carrier for the disorder. Figure 16.5 shows how females can become carriers.

1. **a.** What is the genotype for a color-blind female?_____ How many recessive alleles does a female inherit to be color-blind? _____

   b. What is the genotype for a color-blind male? _____ How many recessive alleles does a male inherit to be color-blind? _____

Figure 16.5 Two common patterns of X-linked inheritance in humans.

a. The sons of a carrier mother have a 50% chance of being color blind. **b.** A color-blind father has carrier daughters.

In these questions, calculate the chances for males and females separately.

2. a. With reference to Figure 16.5*a*, if the mother is a carrier (_____) and the father has normal vision

(_____), what are the chances that a daughter will be color-blind? _____

b. A daughter will be a carrier? _____ **c.** A son will be color-blind? _____

3. a. With reference to Figure 16.5*b*, if the mother has normal vision (_____) and the father is color-blind

(_____), what are the chances that a daughter will be color-blind? _____

b. A daughter will be a carrier? _____ **c.** A son will be color-blind? _____

X-Linked Genetics Problems

For **color blindness,** there are two possible X-linked alleles involved. One affects the green-sensitive cones, whereas the other affects the red-sensitive cones. Overall about 7–10% of males have red-green color blindness, meaning that both alleles are affected.

1. A woman with normal color vision (_____), whose father was color-blind (_____), marries a man

with normal color vision (_____). What genotypes could occur among their offspring? _____

_____

What genotypes could occur if it was the normal-visioned man's father who was color-blind?

_____

2. John's father is color-blind (_____) but his mother is not color-blind (_____).

Is John necessarily color-blind? _____ Explain. _____

Could he be color-blind? _____ Explain. _____

Hemophilia is called the bleeder's disease because the affected person's blood is unable to clot. Although hemophiliacs do bleed externally after an injury, they also suffer from internal bleeding, particularly around joints. Hemorrhages can be checked with transfusions of fresh blood (or plasma) or concentrates of the clotting protein. The most common type of hemophilia is hemophilia A, due to absence or minimal presence of a particular clotting factor called factor VIII.

Inheritance of Multiple Alleles

When a trait is controlled by **multiple alleles,** the gene has more than two possible alleles. But each person has only two of the possible alleles. For example, ABO blood type is determined by multiple alleles: I^A, I^B, i. Red blood cells have surface molecules called antigens that indicate they belong to the person. The I^A allele causes red blood cells to carry an A antigen, the I^B allele causes red blood cells to carry a B antigen, and the i allele causes the red blood cells to have neither of these antigens. I^A and I^B are dominant to i. Remembering that each person can have any two of the possible alleles, these are possible genotypes and phenotypes for blood types.

Genotypes	Antigens on Red Cells	Blood Types
$I^A I^A$, $I^A i$	A	A
$I^B I^B$, $I^B i$	B	B
$I^A I^B$	A and B	AB
ii	none	O

Blood type also indicates whether the person is Rh positve or Rh negative. If the genotype is *DD* or *Dd,* the person is Rh positive and if the genotype is *dd,* the person is Rh negative. It is customary to simply attach a + or − superscript to the ABO blood type, as in A⁻.

Experimental Procedure: Using Blood Type to Help Determine Paternity

In this experimental procedure a mother, Wanda, is seeking support for her child, Sophia. We will use blood typing to decide which of three men could possibly be the father.

1. Obtain three testing plates, each of which contains three depressions; vials of blood from possible fathers 1, 2, and 3, respectively; vials of anti-A serum, anti-B serum, and anti-Rh serum. (All of these are synthetic.)
2. Using a wax pencil, number the plates so you know which plate is for possible father 1, 2, or 3. Look carefully at a plate and notice the wells are designated as A, B, or Rh.
3. Being sure to close the cap to each vial in turn, do the following using plate #1:
   Add a drop of father 1 blood to all three wells—close the cap.
   Add a drop of anti-A (blue) to the well designated A—close the cap.
   Add a drop of anti-B (yellow) to the well designated B—close the cap.
   Add a drop of anti-Rh (clear) to the well designated Rh—close the cap.
4. Stir the contents of each well with a mixing stick of the correct color. After a few minutes, examine the wells for agglutination (i.e., granular appearances that indicate the blood type). (Rh⁺ takes the longest to react.) If a person had AB⁺ blood, which wells would show agglutination? _____

   _____

5. Repeat steps 3 and 4 for plates 2 and 3.
6. Record the blood type results for each of the men in Table 16. 2.

Table 16.2 Blood Types of Involved Persons					
	Mother*	Child*		Father?	
	Wanda	Sophia	#1	#2	#3
Blood type	B⁻	AB⁺			

*Your instructor may have you confirm these results.

Conclusion

1. Noting that only father 3 could have given Sophia the Rh antigen, from whom did she receive the

   I^B allele? _____ From which parent did she receive the I^A allele? _____ Is there any other

   possible interpretation to the results of blood typing? _____

   _____

Blood Typing Problems

1. A man with type A blood reproduces with a woman who has type B blood. Their child has blood type O. Using I^A, I^B, and i, give the genotype of all persons involved.

   man _____ woman _____ child _____

2. If a child has type AB blood and the father has type B blood, what could the genotype of the mother be?

   _____ or _____

3. If both mother and father have type AB blood, they cannot be the parents of a child who has what blood

   type? _____

4. What blood types are possible among the children if the parents are $I^Ai \times I^Bi$? (*Hint:* Do a Punnett square using the possible gametes for each parent.)

   Punnett Square:

16.3 Genetic Counseling

Potential parents are becoming aware that many illnesses are caused by abnormal chromosomal inheritance or by gene mutations. Therefore, they are seeking genetic counseling, which is available in many major hospitals. The counselor helps the couple understand the mode of inheritance for a condition of concern so that the couple can make an informed decision about how to proceed.

Down syndrome

Determining Chromosomal Inheritance

If a genetic counselor suspects that a condition is due to a chromosome anomaly, he or she may suggest that the chromosomal inheritance be examined. It is possible to view the chromosomes of an individual because cells can be microscopically examined and photographed just before cell division occurs. A computer is then used to arrange the chromosomes by pairs. The resulting pattern of chromosomes is called a **karyotype.**

   A trisomy occurs when the individual has three chromosomes instead of two chromosomes at one karyotype location. **Trisomy 21** (Down syndrome) is the most common autosomal trisomy in humans. Survival to adulthood is common. Characteristic facial features include an eyelid fold, a flat face, and a large fissured tongue. Some degree of intellectual impairment is common, as is early-onset Alzheimer disease. Sterility due to sexual underdevelopment may be present.

Down syndrome karyotype.

A female with **Turner syndrome** (XO) has only one sex chromosome, an X chromosome; the O signifies the absence of the second sex chromosome. Because the ovaries never become functional, these females do not undergo puberty or menstruation, and their breasts do not develop. Generally, females with Turner syndrome have a short build, folds of skin on the back of the neck, difficulty recognizing various spatial patterns, and normal intelligence. With hormone supplements, they can lead fairly normal lives.

Turner syndrome   XO   Poly-X syndrome   XXX

When an egg having two X chromosomes is fertilized by an X-bearing sperm, an individual with **poly-X syndrome** results. The body cells have three X chromosomes and therefore 47 chromosomes. Although they tend to have learning disabilities, poly-X females have no apparent physical anomalies, and many are fertile and have children with a normal chromosome count.

When an egg having two X chromosomes is fertilized by a Y-bearing sperm, a male with **Klinefelter syndrome**

Klinefelter syndrome   XXY   Jacob syndrome   XYY

results. This individual is male in general appearance, but the testes are underdeveloped, and the breasts may be enlarged. The limbs of XXY males tend to be longer than average, muscular development is poor, body hair is sparse, and many XXY males have learning disabilities.

Jacob syndrome occurs in males who are usually taller than average, suffer from persistent acne, and tend to have speech and reading problems. At one time, it was suggested that XYY males were likely to be criminally aggressive, but the incidence of such behavior has been shown to be no greater than that among normal XY males.

Label each karyotype in Figure 16.6 and indicate the sex chromosomes for that syndrome. Explain your answers on the lines provided.

Figure 16.6   Sex chromosome anomalies.

a. _____   b. _____   c. _____   d. _____

_____   _____   _____   _____

_____   _____   _____   _____

Determining the Pedigree

A **pedigree** shows the inheritance of a genetic disorder within a family. A pedigree can help determine the inheritance pattern and whether any particular individual has an allele for that disorder. Then a Punnett square can be done to determine the chances of a couple producing an affected child.

In a pedigree, roman numerals indicate the generation, and arabic numerals indicate particular individuals in that generation. The symbols used to indicate normal and affected males and females, reproductive partners, and siblings are shown in Figure 16.7.

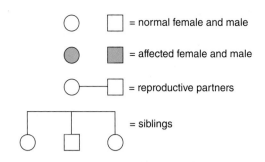

Figure 16.7   Pedigree symbols.

Pedigree Analyses

For each of the following pedigrees, determine how a genetic disorder is inherited. Is the inheritance pattern autosomal dominant, autosomal recessive, or X-linked recessive? Also, decide the genotype of particular individuals in the pedigree. A pedigree indicates the phenotype, and you can reason out the genotype.

1. Study the following pedigree:

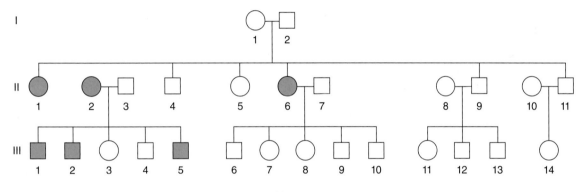

a. What is the inheritance pattern for this genetic disorder? _____

b. What is the genotype of the following individuals? Use *A* for the dominant allele and *a* for the recessive allele.

Generation I, individual 1: _____

Generation II, individual 1: _____

Generation III, individual 8: _____

2. Study the following pedigree:

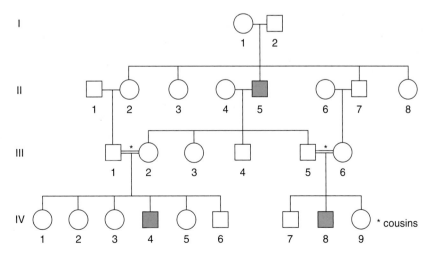

* cousins

a. What is the inheritance pattern for this genetic disorder? _____

b. What is the genotype of the following individuals?

Generation I, individual 1: _____

Generation II, individual 8: _____

Generation III, individual 1: _____

3. Study the following pedigree:

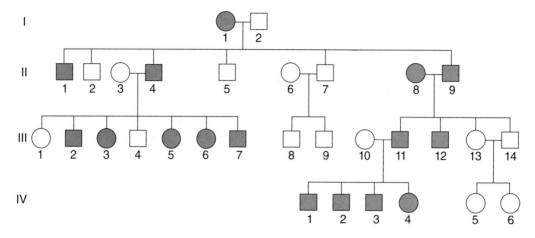

a. What is the inheritance pattern for this genetic disorder? _____

b. What is the genotype of the following individuals?

Generation I, individual 1: _____

Generation II, individual 7: _____

Generation III, individual 11: _____

Construction of a Pedigree

You are a genetic counselor who has been given the following information from which you will construct a pedigree.

Your data: Henry's maternal grandfather and his mother have double eyelashes. Their spouses do not. Henry who has double eyelashes is married to Isabella who has normal eyelashes, and their first child Polly has normal eyelashes. The couple wants to know the chances of any child having a double row of eyelashes.

1. *Using pedigree symbols only, construct two pedigrees.* Begin with the maternal grandfather and grandmother, and end with Polly. Place it on both the left and right.

Pedigree 1   **Pedigree 2**

2. What is your key for this trait?

Key:

_____ normal eyelashes

_____ double row of eyelashes

3. *Try out a pattern of autosomal dominant inheritance* by assigning appropriate genotypes for the dominant pattern of inheritance to each person in pedigree 1. Then *try out a pattern of X-linked dominant* inheritance by assigning appropriate genotypes for the x-linked pattern of inheritance to each person in pedigree 2. Which pattern is correct?

_____

4. Use correct genotypes to show the cross between Henry and Isabella and perform a Punnett square for this cross.

        Henry   Isabella

_____ × _____

5. What are the chances that a child of Henry and Isabella will have double eyelashes? _____

Choosing the Gender of Your Child

The gender of a child depends upon whether an X-bearing sperm or a Y-bearing sperm enters the egg. A new technology that can separate X-bearing and Y-bearing sperm offers prospective parents the opportunity to choose the sex of their child. The results are not perfect. Following artificial insemination, there's about an 85% success rate for a girl and about a 65% success rate for a boy.

Proponents of sex-selection technology argue that there are many instances in which the ability to choose the sex of a child may benefit society. For example, if a mother is a carrier of an X-linked genetic disorder, such as hemophilia or Duchenne muscular distrophy, this would be the simplest way to ensure a healthy child. Previously, a pregnant woman with these concerns had to wait for the results of an amniocentesis test and then decide whether to abort the pregnancy if it was a boy. In such cases, is it better to ensure that the child will not have a specific genetic disorder than to take the risk?

Also, it's possible for parents to use this technology simply because they choose to have a child of one particular gender. Do you think it is acceptable to select the sex of your child? Some are concerned that allowing parents to select the sex of their children could lead to selecting specific traits for their children. Should this be a concern?

_____   1. Mary's father does not have freckles but Mary does. What genotypes could Mary's mother have?

_____   2. What is the genotype of a man who has unattached earlobes but whose mother has attached earlobes?

_____   3. A cross gives a 3:1 phenotypic ratio. What are the genotypes of the parents?

_____   4. What does a geneticist construct to show the inheritance pattern of a genetic disorder within a family?

_____   5. The alleles of which parent, regardless of the phenotype, determine color blindness in a son?

_____   6. If only males are affected in a pedigree chart, what is the likely inheritance pattern for the trait?

_____   7. Which cross gives the better chance of an offspring with the recessive phenotype $Aa \times Aa$ or $Aa \times aa$?

_____   8. Mary has a widow's peak but her sister has a smooth hairline. Is either one of Mary's parents homozygous dominant or recessive?

_____   9. Both parents have attached earlobes (recessive). What percentage of children will have unattached earlobes?

_____   10. Mary (blood type A) has a child who is blood type O. What is Mary's genotype?

_____   11. A woman is a carrier for hemophilia. What are the chances for sons with hemophilia if the father does not have hemophilia?

_____   12. If the parents are not affected and a child is affected, what is the inheritance pattern?

_____   13. How do you recognize a karyotype of a person with Klinefelter syndrome?

_____   14. A man has Huntington disease (autosomal dominant). He cannot assume his mother passed him the gene. Why not?

_____   15. A boy is a hemophiliac but his mother is not. What is her genotype?

_____   16. The trait is autosomal recessive and the results of a cross are 1:1. Using A = dominant and a = recessive, give the genotypes of the parents.

_____   17. Give the genotype of a girl that is a hemophiliac.

Thought Questions

18. What would be the genotype of a man who is homozygous dominant for widow's peak and is color blind. (_Hint:_ First do the genotype for widow's peak, and then do the genotype for color blindness.)

19. What is the probability a woman heterozygous for an X-linked trait will have a son with a genetic disorder if the genetic disorder is recessive? If the genetic disorder is dominant? Explain.

20. If a pedigree shows that the daughters of affected fathers (the mother is homozygous normal) are affected, what could be the pattern of inheritance if the offending allele is on the X chromosome?

McGraw-Hill Access Science Website

An online encyclopedia of science and technology that provides information, including videos, that can enhance the laboratory experience.

www.accessscience.com

The following LearnSmart Lab contains exercises that are related to the content of this chapter:

• Human Genetics

Find even more resources to learn the chapter concepts including animations, tutorial videos, and interactive practice questions.

17

DNA and Biotechnology

Learning Outcomes

17.1 DNA Structure and Replication
- Explain how the structure of DNA facilitates replication.
- Explain how DNA replication is semiconservative.

Prelab Question: What does *semiconservative replication* mean?

17.2 RNA Structure
- Compare DNA and RNA, discussing their similarities as well as what distinguishes one from the other.

Prelab Question: The "R" in RNA stands for what?

17.3 DNA and Protein Synthesis
- Describe how DNA is able to store the information that specifies a protein.
- Compare the events of transcription with those of translation during protein synthesis.

Prelab Question: Which one, transcription or translation, produces a protein?

17.4 Isolation of DNA and Biotechnology
- Describe the procedure for isolating DNA in a test tube.
- Understand the process of DNA gel electrophoresis.

Prelab Question: Give an example to show that DNA can be manipulated in the laboratory like any other chemical.

17.5 Detecting Genetic Disorders
- Understand the relationship between an abnormal DNA base sequence and a genetic disorder.
- Suggest two ways to detect a genetic disorder.

Prelab Question: If the DNA base sequence changes, the protein changes. How?

Application for Daily Living: Personal DNA Sequencing

Introduction

This laboratory pertains to molecular genetics and biotechnology. Molecular genetics is the study of the structure and function of **DNA (deoxyribonucleic acid)**, the genetic material. **Biotechnology** is the manipulation of DNA for the benefit of human beings and other organisms. Significant advances in medicine, agriculture, and science in general can be attributed to the fields of molecular genetics and biotechnology.

First we will study the structure of DNA and see how that structure facilitates DNA replication in the nucleus of cells. DNA replicates prior to cell division; following cell division, each daughter cell has a complete copy of the genetic material. DNA replication is also needed to pass genetic material from one generation to the next. You may have an opportunity to use models to see how replication occurs.

Then we will study the structure of **RNA (ribonucleic acid)** and how it differs from that of DNA, before examining how DNA and RNA specify protein synthesis. The linear construction of DNA, in which nucleotide is linked to nucleotide, is paralleled by the linear construction of the primary structure of protein, in which amino acid is linked to amino acid. Essentially, we will see that the sequence of nucleotides in DNA codes for the sequence of amino acids in a protein. We will also review the role of three types of RNA in protein synthesis. DNA's code is passed to messenger RNA (mRNA), which moves to the ribosomes containing ribosomal RNA (rRNA). Transfer RNA (tRNA) brings the amino acids to the ribosomes.

We now understand that a mutated gene has an altered DNA base sequence, which can lead to a genetic disorder. You will have an opportunity to carry out a laboratory procedure that detects whether an individual is normal, has sickle-cell disease, or is a carrier.

Know Thyamine

17.1 DNA Structure and Replication

The structure of DNA lends itself to **replication,** the process that makes a copy of a DNA molecule. Accurate DNA replication is a necessary part of chromosome duplication, which precedes cell division. It also makes possible the passage of DNA from one generation to the next.

DNA Structure

DNA is a polymer of nucleotide subunits (Fig. 17.1). Each nucleotide is composed of three molecules: deoxyribose (a 5-carbon sugar), a phosphate, and a nitrogen-containing base, and is attached to its complementary nucleotide by hydrogen bonds.

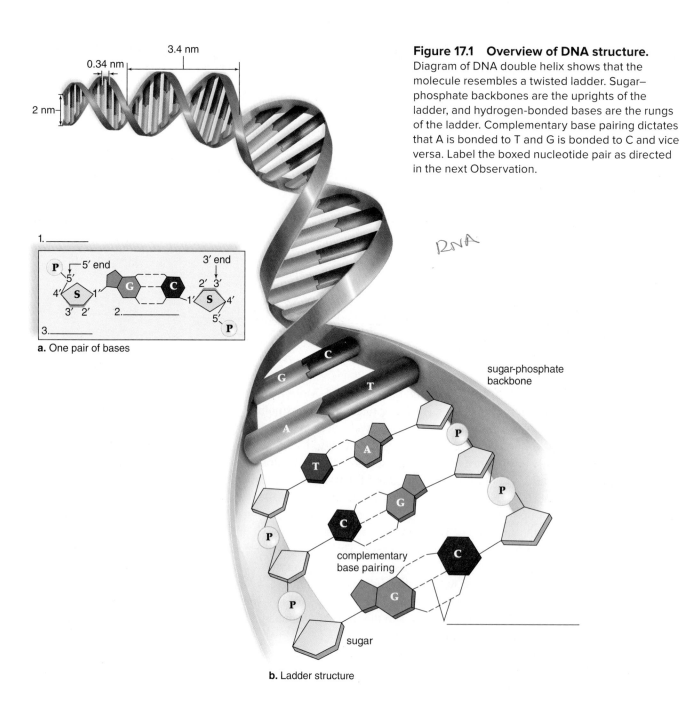

Figure 17.1 Overview of DNA structure.
Diagram of DNA double helix shows that the molecule resembles a twisted ladder. Sugar–phosphate backbones are the uprights of the ladder, and hydrogen-bonded bases are the rungs of the ladder. Complementary base pairing dictates that A is bonded to T and G is bonded to C and vice versa. Label the boxed nucleotide pair as directed in the next Observation.

RNA

a. One pair of bases

sugar-phosphate backbone

complementary base pairing

sugar

b. Ladder structure

1. A boxed nucleotide pair is shown in Figure 17.1*a*. If you are working with a kit, draw a representation of one of your nucleotides here. *Label phosphate, base pair, and deoxyribose in your drawing and in Figure 17.1*a.

2. Notice the four types of bases: cytosine (C), thymine (T), adenine (A), and guanine (G). What is the color of the four types of bases in Figure 17.1? In your kit? Complete Table 17.1 by writing in the colors of the bases.

Table 17.1   Base Colors

	In Figure 17.1	In Your Kit
Cytosine		
Thymine		
Adenine		
Guanine		

3. Using Figure 17.1 as a guide, join several nucleotides together. Observe the entire DNA molecule. What types of molecules make up the backbone of DNA (Fig. 17.1*b*)? _____ and _____

   In the backbone, the phosphate of one nucleotide is bonded to a sugar of the next nucleotide by a covalent bond.

4. Using Figure 17.1 as a guide, join the bases together with hydrogen bonds. *Label a hydrogen bond in Figure 17.1*b. Dashes are used to represent hydrogen bonds in Figure 17.1 because hydrogen bonds are (strong or weak) _____.

5. In Figure 17.1*b* and in your model, the base A is always paired with the base _____, and the base C is always paired with the base _____. This is called complementary base pairing.

6. The DNA molecule resembles a twisted ladder. The sugar-phosphate backbones equate to the uprights of the ladder. What molecules equate to the rungs of the ladder?

   _____

7. Each half of the DNA molecule is a DNA strand. Why is DNA also called a double helix (Fig. 17.1*b*)?

   _____

   _____

DNA Replication

During replication, the DNA molecule is duplicated so that there are two DNA molecules. We will see that complementary base pairing makes replication possible.

Observation: DNA Replication

1. Before replication begins, DNA is unzipped. Using Figure 17.2a as a guide, break apart your two DNA strands. What bonds are broken to unzip the DNA strands?

   _____

2. Using Figure 17.2b as a guide, attach new complementary nucleotides to each strand using complementary base pairing.

3. Complete Table 17.2 to show that you understand complementary base pairing. You now have two DNA molecules (Fig. 17.2c). Are your molecules identical?

   _____

4. Because of complementary base pairing, each new double helix is

   composed of an _____

   strand and a _____ strand.

   Write *old* or *new* beside each strand in Figure 17.2a, b, and c, 1–10. *Conservative* means to save something from the past. Why is DNA replication called semiconservative?

   _____
   _____
   _____
   _____

Figure 17.2   DNA replication.
Use of the ladder configuration better illustrates how replication takes place. **a.** The parent DNA molecule. **b.** The "old" strands of the parent DNA molecule have separated. New complementary nucleotides available in the cell are pairing with those of each old strand. **c.** Replication is complete.

1. _____   2. _____

a.

3. _____   4. _____

5. _____   6. _____

b.

7. _____   8. _____

9. _____   10. _____

c.

Table 17.2 DNA Replication	
Old strand	G G G T T C C A T T A A A T T C C A G A A A T C A T A
New strand	

5. Genetic material has to be inherited from cell to cell and organism to organism. Consider that because of DNA replication, a chromosome is composed of two chromatids, and each chromatid is a complete DNA molecule. The chromatids separate during cell division so that each daughter cell receives a copy of each chromosome. Does replication provide a means for passing DNA from cell to cell and organism to organism? _____ Explain. _____

_____

17.2 RNA Structure

Like DNA, RNA is a polymer of nucleotides (Fig. 17.3). In an RNA nucleotide, the sugar ribose is attached to a phosphate molecule and to a nitrogen-containing base, C, U, A, or G. In RNA, the base uracil replaces thymine as one of the pyrimidine bases. RNA is single stranded, whereas DNA is double stranded.

Figure 17.3 Overview of RNA structure.
RNA is a single strand of nucleotides. Label the boxed nucleotide as directed in the next Observation.

1. Describe the backbone of an RNA molecule. _____
2. Where are the bases located in an RNA molecule? _____
3. Complete Table 17.3 to show the complementary DNA bases for the RNA bases.

Table 17.3 DNA and RNA Bases			
RNA Bases	C	U	A	G
DNA Bases				

1. If you are using a kit, draw a nucleotide for the construction of mRNA. *Label the ribose (the sugar in RNA), the phosphate, and the base in your drawing and in Figure 17.3, page 235.*

2. Complete Table 17.4 by writing in the colors of the bases in Figure 17.3 and in your kit.

Table 17.4 Base Colors		
	In Figure 17.3	In Your Kit
Cytosine		
Uracil		
Adenine		
Guanine		

3. The base uracil substitutes for the base thymine in RNA. Complete Table 17.5 to show several other ways RNA differs from DNA.

Table 17.5 DNA Structure Compared with RNA Structure		
	DNA	RNA
Sugar	Deoxyribose	
Bases	Adenine, guanine, thymine, cytosine	
Strands	Double stranded with base pairing	
Helix	Yes	

17.3 DNA and Protein Synthesis

Protein synthesis requires the processes of transcription and translation. During **transcription,** which takes place in the nucleus, an RNA molecule called **messenger RNA (mRNA)** is made complementary to one of the DNA strands. This mRNA leaves the nucleus and goes to the ribosomes in the cytoplasm. Ribosomes are composed of **ribosomal RNA (rRNA)** and proteins in two subunits.

During **translation,** RNA molecules called **transfer RNA (tRNA)** bring amino acids to the ribosome, and they join in the order prescribed by mRNA. In the end, the final sequence of amino acids in a protein is specified by DNA. This is the information that DNA, the genetic material, stores. What is the role of each of these participants in protein synthesis?

DNA _____

mRNA _____

tRNA _____

Transcription

During transcription, complementary RNA is made from a DNA template (Fig. 17.4). A portion of DNA unwinds and unzips at the point of attachment of RNA polymerase. A strand of mRNA is produced when complementary nucleotides join in the order dictated by the sequence of bases in DNA. Transcription occurs in the nucleus, and the mRNA passes out of the nucleus to enter the cytoplasm.

Label Figure 17.4. For number 1, note the name of the enzyme that carries out mRNA synthesis. For number 2, note the name of the entire molecule.

Observation: Transcription

1. If you are using a kit, unzip your DNA model so that only one strand remains. This strand is the **template strand,** the strand transcribed or copied into complementary base pairs of RNA.
2. Using Figure 17.4 as a guide, construct a messenger RNA (mRNA) molecule by first lining up RNA nucleotides complementary to the template strand of your DNA molecule. Join the nucleotides together to form mRNA.
3. A portion of DNA has the sequence of bases shown in Table 17.6. Complete Table 17.6 to show the sequence of bases in mRNA.
4. If you are using a kit, unzip the mRNA transcript from the DNA. Locate the end of the strand that will move to the

   _____ in the cytoplasm.

Figure 17.4   Messenger RNA (mRNA).
Messenger RNA complementary to a section of DNA forms during transcription.

1. _____

DNA template strand

2. _____

Table 17.6	Transcription
DNA	T A C A C G A G C A A C T A A C A T
mRNA	

Translation

DNA specifies the sequence of amino acids in a polypeptide because every three bases stand for an amino acid. Therefore, DNA is said to have a **triplet code.** The bases in mRNA are complementary to those in DNA, and therefore every three bases in mRNA (called a **codon**) stand for the same sequence of amino acids as does DNA. The correct sequence of amino acids in a polypeptide is the message that mRNA carries.

Messenger RNA leaves the nucleus and proceeds to the ribosomes, where protein synthesis occurs. Transfer RNA (tRNA) molecules are so named because they transfer amino acids to the ribosomes. Each tRNA has a specific amino acid at one end and a matching **anticodon** at the other end (Fig. 17.5). The anticodon is complementary to the codon for this amino acid, as we shall soon see. *Label Figure 17.5, where the amino acid is represented as a colored ball, the tRNA is green, and the anticodon is the sequence of three bases.*

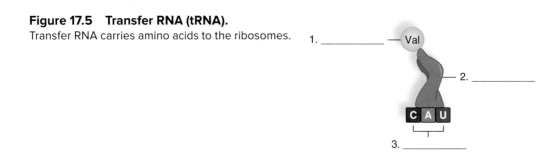

Figure 17.5   Transfer RNA (tRNA).
Transfer RNA carries amino acids to the ribosomes.

Observation: Translation

1. Figure 17.6 shows seven tRNA–amino acid complexes. Every amino acid has a name; in the figure, only the first three letters of the name are inside the ball. *Using the mRNA sequence given in Table 17.7, number the tRNA–amino acid complexes in the order they will come to the ribosome.*
2. If you are using a kit, arrange your tRNA–amino acid complexes in the proper order. *Complete Table 17.7.* Why are the codons and anticodons in groups of three? _____

Figure 17.6   Transfer RNA diversity.
Each type of tRNA carries only one particular amino acid, designated here by the first three letters of its name.

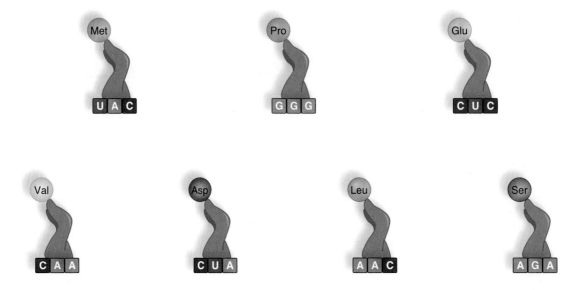

Table 17.7 Translation

mRNA codons	AUG	CCC	GAU	GUU	GAG	UUG	UCU
tRNA anticodons							
Amino acid*							

*Use three letters only. See Table 17.8 for the full names of these amino acids.

Table 17.8 Names of Amino Acids

Abbreviation	Name
Met	methionine
Pro	proline
Asp	aspartic acid
Val	valine
Glu	glutamic acid
Leu	leucine
Ser	serine

3. Figure 17.7 shows the manner in which the polypeptide grows. A ribosome has room for two tRNA complexes at a time. As the first tRNA leaves, it passes its amino acid or peptide to the next tRNA–amino acid complex. Then the ribosome moves forward, making room for the next tRNA–amino acid complex. This sequence of events occurs over and over until the entire polypeptide is borne by the last tRNA to come to the ribosome. *In Figure 17.7, label the ribosome, the mRNA, and the peptide chain.*

Figure 17.7 Protein synthesis.

1. A ribosome has room for two tRNA–amino acid complexes. **2.** Before tRNA leaves, an RNA passes its attached peptide to its neighboring tRNA–amino acid complex. **3.** The ribosome moves forward, and the next tRNA–amino acid complex arrives.

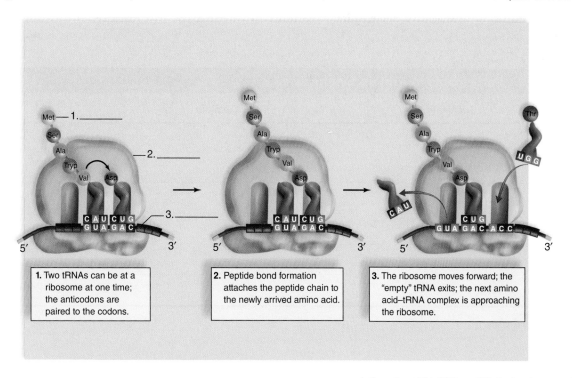

1. Two tRNAs can be at a ribosome at one time; the anticodons are paired to the codons.

2. Peptide bond formation attaches the peptide chain to the newly arrived amino acid.

3. The ribosome moves forward; the "empty" tRNA exits; the next amino acid–tRNA complex is approaching the ribosome.

Summary of Protein Synthesis

Examine Figure 17.8 and complete Table 17.9 to show that you understand the role of the particpants in protein synthesis.

Figure 17.8 Participants in synthesis.

The two steps required for protein synthesis are transcription, which occurs in the nucleus, and translation, which occurs in the cytoplasm at the ribosomes.

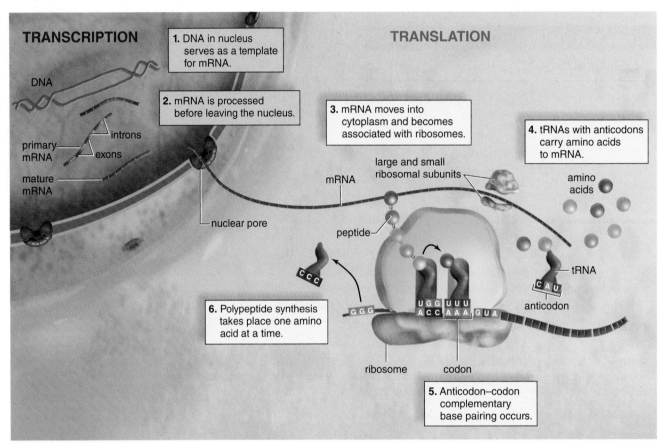

Table 17.9 Participants in Protein Synthesis		
	Special Role	**Explain**
DNA		
mRNA		
tRNA		
rRNA		
Amino acid		
Protein		

17.4 Isolation of DNA and Biotechnology

In the following experiment, you will isolate DNA from the cells of a fruit or vegetable. It will only be necessary to expose the cells to an agent (dishwashing detergent) that emulsifies membrane in order to "free" the DNA from its enclosures (plasma membrane and nuclear envelope). When transferred to a tube, the presence of NaCl allows DNA to precipitate as a sodium salt. The precipitate forms at the interface between ethanol and the salt solution, and then it floats to the top of the tube where it may be collected.

Experimental Procedure: Isolation of DNA

1. You will need a slice of fruit or vegetable (i.e., tomato, onion, or apple), a mortar and pestle, and a large clean glass test tube on ice.

2. Crush and grind a slice of fruit or vegetable in a mortar and pestle. Remove the pestle and set aside. Add enough 0.9% NaCl solution to achieve a "soupy" consistency.

> ⚠ For experimental procedures, wear safety goggles, gloves, and protective clothing. If any chemical spills on your skin, wash immediately with mild soap and water; flood eyes with water only. Report any spills immediately to your instructor.

3. Add two drops of dishwashing detergent (such as blue Dawn) to the mixture in the mortar. Swirl until the color of the detergent disappears. Wait 3 to 5 minutes. The solution becomes clear and viscous as the DNA escapes its enclosures.

4. Check the viscosity of the solution in the mortar, and if it is too viscous for you to pipet, add a few more ml of NaCl solution. Use a transfer pipet to move 2 to 3 ml of the DNA from the mortar to the glass tube. Try to pick up clear zones of the solution WITHOUT CELL DEBRIS.

5. Slowly add 5 to 7 ml of ice-cold ethanol down the side of the tube on ice. As you do so, DNA will precipitate *between* the ethanol and the NaCl solutions. Then it will float to the top, where it can be picked up by a fresh transfer pipet. Any "white dust" you see consists of RNA molecules.

6. Use a transfer pipet to place the DNA in a small, clean test tube. Pipet away any extra water/ethanol and air-dry the DNA for a few minutes. Dissolve the DNA in 3 to 4 ml distilled H_2O.

7. Add five drops of phenol red, a pH indicator. The resulting dark pink color confirms the presence of nucleic acid (i.e., the DNA).

Experimental Procedure: Gel Electrophoresis

During gel electrophoresis, charged DNA molecules migrate across a span of gel (gelatinous slab) because they are placed in a powerful electrical field. In the present experiment, each DNA sample is placed in a small depression in the gel called a well. The gel is placed in a powerful electrical field. The electricity causes DNA fragments, which are negatively charged, to move through the gel according to their size.

> ⚠ **Gel electrophoresis** Students should wear personal protective equipment. Safety goggles and smocks or aprons while loading gels and during electrophoresis and protective gloves while staining.

Figure 17.9 Equipment and procedure for gel electrophoresis.

a. Agarose solution poured into casting tray

b. Comb that forms wells for samples

c. Wells that can be loaded with samples

d. Electrophoresis chamber and power supply

Almost all DNA gel electrophoresis is carried out using horizontal gel slabs (Fig. 17.9). First, the gel is poured onto a plastic plate, and the wells are formed. After the samples are added to the wells, the gel and the plastic plate are put into an electrophoresis chamber, and buffer is added. The DNA samples begin to migrate after the electrical current is turned on. With staining, the DNA fragments appear as a series of bands spread from one end of the gel to the other according to their size because smaller fragments move faster than larger fragments.

Answer the following questions:

a. What is biotechnology? (See page 231.) _____

_____

_____

b. Speculate how the ability to isolate DNA and run gel electrophoresis of DNA relates to biotechnology.

_____

_____

_____

c. Name a biotechnology product someone you know is now using or taking as a medicine. (Examples include insulin, vaccines, ingredients of ice cream, cheese, make-up, detergents, dye for blue jeans, etc.)

_____

17.5 Detecting Genetic Disorders

The base sequence of DNA in all the chromosomes is an organism's genome. Now that the Human Genome Project is finished, we know the usual order of all the 3.6 billion nucleotide bases in the human genome. Someday it will be possible to sequence anyone's genome within a relatively short time, and thereby determine what particular base sequence alterations signify that he or she has a disorder or will have one in the future. In this laboratory, you will study the alteration in base sequence that causes a person to have sickle-cell disease.

In persons with sickle-cell disease, the red blood cells aren't biconcave disks like normal red blood cells—they are sickle-shaped. Sickle-shaped cells can't pass along narrow capillary passageways. They clog the vessels and break down, causing the person to suffer from poor circulation, anemia, and poor resistance to infection. Internal hemorrhaging leads to further complications, such as jaundice, episodic pain in the abdomen and joints, and damage to internal organs.

Sickle-shaped red blood cells are caused by an abnormal hemoglobin (Hb^S). Individuals with the $Hb^A Hb^A$ genotype are normal; those with the $Hb^S Hb^S$ genotype have sickle-cell disease, and those with the $Hb^A Hb^S$ have sickle-cell trait. Persons with sickle-cell trait do not usually have sickle-shaped cells unless they experience dehydration or mild oxygen deprivation.

Genetic Sequence for Sickle-Cell Disease

Examine Figure 17.10, *a* and *b,* which shows the DNA base sequence, the mRNA codons, and the amino acid sequence for a portion of the gene for Hb^A and the same portion for Hb^S. Today many genetic disorders can be detected by genomic sequencing.

1. In what one base does Hb^A differ from Hb^S? Hb^A _____ Hb^S _____

2. What are the codons that contain this base? Hb^A _____ Hb^S _____

3. What is the amino acid difference? Hb^A _____ Hb^S _____

Figure 17.10
Sickle-cell disease.
Sickle-cell disease occurs when (**a**) the DNA base sequence in one location has changed from CTC (**b**) to CAC in both alleles for Hb^A.

DNA   TGA'GGA'CTC'CTC'TTC

transcription

mRNA   ACU'CCU'GAG'GAG'AAG

translation

polypeptide   Thr   Pro   Glu   Glu   Lys

DNA   TGA'GGA'CAC'CTC'TTC

transcription

mRNA   ACU'CCU'GUG'GAG'AAG

translation

polypeptide   Thr   Pro   Val   Glu   Lys

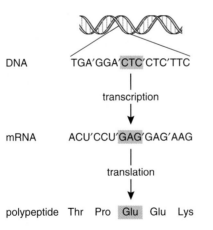

a. Normal red blood cells

b. Sickle-shapedl red blood cell

This amino acid difference causes the polypeptide chain in sickle-cell hemoglobin to pile up as firm rods that push against the plasma membrane and deform the red blood cell into a sickle shape:

glutamatic acid
(polar R group)

valine
(nonpolar R group)

Detection of Sickle-Cell Disease by Gel Electrophoresis

Three samples of hemoglobin (A, B, and C) have been subjected to protein gel electrophoresis. Protein gel electrophoresis is carried out in the in the same manner as DNA gel electrophoresis (Fig. 17.9) except the gel has a different composition.

1. Sickle-cell hemoglobin (Hb^S) migrates slower toward the positive pole than normal hemoglobin (Hb^A) because the amino acid valine has no polar R groups, whereas the amino acid glutamate does have a polar R group.

2. In Figure 17.11, which lane contains only Hb^S, signifying that the individual is $Hb^S Hb^S$? _____

3. Which lane contains only Hb^A, signifying that the individual is $Hb^A Hb^A$? _____

4. Which lane contains both Hb^S and Hb^A, signifying that the individual is $Hb^A Hb^S$? _____

Figure 17.11   Gel electrophoresis of hemoglobins.

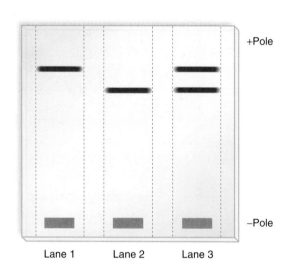

+Pole

−Pole

Lane 1   Lane 2   Lane 3

Detection by Genomic Sequencing

You are a genetic counselor. A young couple seeks your advice because sickle-cell disease occurs among the family members of each. You order DNA base sequencing to be done. The results come back that at one of the loci for normal hemoglobin, each has the abnormal sequence CAC instead of CTC. The other locus is normal. What are the chances that this couple will have a child with sickle-cell disease? _____

Conclusion: Detecting Genetic Disorders

- What two methods of detecting sickle-cell disease were described in this section? _____

  _____

- Which method is more direct and probably requires more expensive equipment to do? _____

  _____

- Which method probably preceded the other method as a means to detect sickle-cell disease? _____

  _____

Application for Daily Living

Personal DNA Sequencing

You may know that it took many years to get the human genome sequenced, but now personal DNA base sequencing can be done within days, and it is relatively affordable. The information can allow a trained geneticist to predict which genetic disorders you are susceptible to, and then you can choose either prevention and/or early detection to avoid a serious illness. DNA sequencing may also tell you which drugs might be especially useful to you in case of illness.

The move is on to require people to go through their personal physician to get their DNA sequenced. Why? It is always a good idea to have it done by a reputable company and to have someone help you sift through the data that might be made available. It's not always easy to learn that you may come down with a disorder, and you would want to know what your options are right away.

Laboratory Review 17

_____ 1. The DNA structure resembles a twisted ladder. What molecules make up the uprights of the ladder?

_____ 2. What makes up the rungs of the ladder?

_____ 3. Do the two DNA double helices following DNA replication have the same, or a different, composition?

_____ 4. If the adenine bases account for 20% of the DNA bases, what would be the percentage of thymine?

_____ 5. If the codons are AUG, CGC, and UAC, what are the anticodons?

_____ 6. Where does protein synthesis take place?

_____ 7. During transcription, what type of RNA is formed that carries the codons?

_____ 8. In what part of the cell does translation occur?

_____ 9. During translation, what type of RNA carries amino acids to the ribosomes?

_____ 10. A person with genotype $Hb^S Hb^S$ has what genetic disorder?

_____ 11. Sickle-cell disease illustrates that a mutation is a change in what?

_____ 12. During gel-electrophoresis normal hemoglobin separates from sickle-cell hemoglobin because of a _____ difference.

_____ 13. What procedure allows investigations to determine the present or future genetic disorders of an individual?

Thought Questions

14. What role does mRNA play in transcription and translation?

15. Explain the manner in which gel electrophoresis identifies a person with sickle-cell disease.

16. Below is a sequence of bases associated with the template DNA strand:

TAC CCC GAG CTT

    a. Identify the sequence of bases in the mRNA resulting from the transcription of the above DNA sequence.

    b. Identify the sequence of bases in the tRNA anticodon that will bind with the first codon of the mRNA identified above.

McGraw-Hill Access Science Website

An online encyclopedia of science and technology that provides information, including videos, that can enhance the laboratory experience.

www.accessscience.com

Find even more resources to learn the chapter concepts including animations, tutorial videos, and interactive practice questions.

LEARNSMART

The following LearnSmart Lab contains exercises that are related to the content of this chapter:

• DNA Biology and Biotechnology

18

Human Evolution

Learning Outcomes

18.1 Fossil Record

- Use the geologic timescale to trace the evolution of humans in broad outline.
- Describe several types of fossils and explain how fossils help to establish the sequence in the evolution of life.
- Explain how scientists use fossils to establish relationships between animals and other forms of life.

Prelab Question: Explain in what way fossils provide evidence that life has a history written in the rocks.

18.2 Comparative Anatomy

- Explain how comparative anatomy provides evidence that humans are related to other groups of vertebrates and also other groups of primates.
- Compare hominid skulls and hypothesize a possible evolutionary sequence.
- Compare the human skeleton with the chimpanzee skeleton, and explain the differences on the basis of their different ways of life.

Prelab Question: What anatomical features show that a chimpanzee skeleton and a human skeleton share a unity of structure?

18.3 Molecular Evidence

- Explain how molecular data aids the study of how humans are related to all other groups of organisms on Earth.
- Explain how molecular data also helps show the degree to which humans are related to other vertebrates and primates.

Prelab Question: If a DNA sequence is the same between chimpanzees and humans, what other type sequence will be the same?

Introduction

As you know, science relies on observation and experimentation to come to conclusions—and one of these conclusions is that human beings are related through **evolution** to all other organisms on Earth. What is evolution? The process is amazingly simple: a group of organisms change over time because the members of the group most suited to the natural environment have more offspring than the others in the group. So, for example, among bacteria those who can withstand an antibiotic leave more offspring and with time the entire group of bacteria becomes resistant to the antibiotic. Reproduction, and therefore evolution, have been going on since the first cells appeared on Earth, and through studying the (1) fossil record, (2) comparative anatomy both anatomical and embryological, and (3) molecular evidence, science is able to tell us the pathway by which humans evolved. Figure 18.1 traces this pathway for you. As shown in Figure 18.1a, the major groups of organisms we see about us today evolved from the first cells. The first animals were simply blobs that survived by feeding on the other groups for nourishment. These first animals gave rise to all the other groups of animals, among them the vertebrates, animals with a backbone (Fig. 18.1b). The vertebrates gave rise to other groups, among them the mammals such as the primates who lived in trees (Fig. 18.1c). The primates evolved, and among them were humans (Fig. 18.1d).

In this lab we will look at the evidence that scientists present to us so that we can also conclude that human beings are related through evolution to all other organisms on Earth.

a. Evolution of Life

b. Evolution of Animals

c. Evolution of Vertebrates

d. Evolution of Primates

Figure 18.1   Evolution of humans.
Human beings are related to all the other types of organisms on Earth. **a.** Humans are related to the other major groups of organisms: bacteria, protists (e.g., Euglena), fungi (e.g., mushrooms), plants, and all other animals. **b.** As animals, they are related to all other animals, such as worms, clams, lobsters, and starfish, and especially other chordates (includes vertebrates). **c.** As vertebrates, they are related to fishes, frogs, alligators, birds, and especially the other mammals. **d.** As mammals, they are most closely related to other primates, especially the apes.

18.1   Fossil Record

All life-forms evolved from the first cell or cells, so life has a history, and this history is revealed by the fossil record. A **fossil** is any evidence of the existence of an organism in ancient times as opposed to modern times. The geologic timescale, developed by geologists and paleontologists, depicts the history of life based on the fossil record (see Table 18.1).

Era	Period	Epoch	Million Years Ago (MYA)	Plant Life	Animal Life
Recent Time		Holocene	(0.01–0)	Human influence on plant life	Age of *Homo sapiens*
	Quaternary			**Significant Mammalian Extinction**	
		Pleistocene	(1.80–0.01)	Herbaceous plants spread and diversify.	Presence of Ice Age mammals. Modern humans appear.
Cenozoic		Pliocene	(5.33–1.80)	Herbaceous angiosperms flourish.	First hominids appear.
		Miocene	(23.03–5.33)	Grasslands spread as forests contract.	Apelike mammals and grazing mammals flourish; insects flourish.
	Tertiary	Oligocene	(33.9–23.03)	Many modern families of flowering plants evolve.	Browsing mammals and monkeylike primates appear.
		Eocene	(55.8–33.9)	Subtropical forests with heavy rainfall thrive.	All modern orders of mammals are represented.
		Paleocene	(65.5–55.8)	Flowering plants continue to diversify.	Primitive primates, mammalian herbivores, carnivores, and insectivores appear.
				Mass Extinction: Dinosaurs and Most Reptiles	
	Cretaceous		(145.5–65.5)	Flowering plants spread; conifers persist.	Placental mammals appear; modern insect groups appear.
Mesozoic	Jurassic		(199.6–145.5)	Flowering plants appear.	Dinosaurs flourish; birds appear.
				Mass Extinction	
	Triassic		(251–199.6)	Forests of conifers and cycads	First mammals appear; first dinosaurs appear; corals and molluscs dominate seas.
				Mass Extinction	
	Permian		(299–251)	Gymnosperms diversify.	Reptiles diversify; amphibians decline.
	Carboniferous		(359.2–299)	Age of great coal-forming forests; tree ferns, club mosses, and horsetails flourish.	Amphibians diversify; first reptiles appear; first great radiation of insects.
Paleozoic (cont'd)				**Mass Extinction**	

Era	Period	Million Years Ago (MYA)	Plant Life	Animal Life
Paleozoic	Devonian	(416–359.2)	First seed plants appear. Seedless vascular plants diversify.	First insects and first amphibians appear on land.
	Silurian	(443.7–416)	Seedless vascular plants appear.	Jawed fishes diversify and dominate the seas.
			Mass Extinction	
	Ordovician	(488.3–443.7)	Nonvascular plants appear on land.	First jawless and then jawed fishes appear.
	Cambrian	(542–488.3)	Marine algae flourish.	All invertebrate phyla present; first chordates appear.
Precambrian		630	Soft-bodied invertebrates	
		1,000	Protists evolve and diversify.	
		2,200	First eukaryotic cells	
		2,700	O_2 accumulates in atmosphere.	
		3,800	First prokaryotic cells	
Distant Time		4,570	Earth forms.	

Geologic Timescale

Divisions of the Timescale

The timescale divides the history of Earth into eras, then periods, and then epochs. The three **eras** span the greatest amounts of time, and the **epochs** are the shortest time frames. Only the **periods** of the Cenozoic era are divided into epochs, meaning that more attention is given to the evolution of primates and flowering plants than to the earlier evolving organisms. Modern civilization is given its own epoch, despite the fact that humans have only been around about .04% of the history of life.

1. List the four eras in the timescale starting with Precambrian time. _____

_____

2. During the _____ era and the _____ period, the first flowering plants appear. How many million years ago was this? _____

3. Among primates, humans and the great apes are hominids. During what period and epoch did hominids appear? _____

Dating Within the Timescale

The timescale provides both relative dates and absolute dates. When you say, for example, "Flowering plants evolved during the Jurassic period," you are using relative time, because flowering plants evolved earlier or later than groups in other periods. If you use the dates given in millions of years, you are using absolute time. Absolute dates are usually obtained by measuring the amount of a radioactive isotope in the rocks surrounding the fossils. Why wouldn't you expect to find human fossils and dinosaur fossils together in rocks of the same date? _____

_____

Fossils

The fossil record depends heavily on anatomical data to show the evolutionary changes that have occurred

as one group gives rise to another group. Why would that be? _____

_____

Invertebrate Fossils

The various groups of **invertebrates** (animals without a backbone) are shown in Figure 18.1b. Even the chordates which is the phylum that contains the **vertebrates** (animals with a backbone) contain a couple of insignificant groups of invertebrates.

In order to familiarize yourself with possible invertebrate fossils, a few of the most common fossilized invertebrate groups are depicted in Figure 18.2.

Figure 18.2 Invertebrate fossils.

Fossilized ammonite, a mollusk

Fossilized brittle stars, echinoderms

Fossilized mantis in amber, an arthropod

Fossilized pseudoscorpion, an arthropod

Fossilized snails, mollusks

Fossilized trilobites, arthropods

Observation: Invertebrate Fossils

1. Obtain a box of selected fossils. If the fossils are embedded in rocks, examine the rock until you have found the fossil. Fossils are embedded in rocks because the sediment that originally surrounded them hardened over time. Most fossils consist of hard parts such as shells, bones, or teeth because these parts are not consumed

or decomposed over time. One possible reason the Cambrian might suddenly be rich in fossils is because organisms before this time did not have _____

_____.

2. The kit you are using, or your instructor, will identify which of the fossils are invertebrate animals. List the types of fossils in Table 18.2, and give a description of the hard part of the fossil that survived. Use the Geologic Timescale (Table 18.1) to tell the era and period they were most abundant; list the fossils from the latest (top) to earliest (bottom).

Vertebrate Fossils

The various groups of vertebrates are shown in Figure 18.1*c*. Today it is generally agreed that birds are reptiles rather than being a separate group. That means that the major groups of vertebrates are (1) various types of fishes (jawless, jawed, cartilaginous, and bony fishes), (2) amphibians such as frogs and salamanders, (3) reptiles such as lizards and crocodiles, and (4) mammals. There are many types of mammals, from whales to mice to humans. The fossil record can rely on skeletal differences such as limb structure to tell if an animal is a mammal.

To help you familiarize yourself with possible invertebrate fossils, a few of the most common fossilized invertebrate groups are depicted in Figure 18.3.

Table 18.2	Invertebrate Fossils		
Type of Fossil		**Era, Period**	**Description of Hard Part**

Figure 18.3 Vertebrate fossils.

Fossilized bird, a reptile

Fossilized deerlike mammal

Fossilized duckbill dinosaur, a reptile

Fossilized bony fish

Fossilized frog, an amphibian

Fossilized snake, a reptile

Table 18.3 Vertebrate Fossils

Type of Fossil	Era, Period	Description of Hard Part

Observation: Vertebrate Fossils

Which of the fossils available to you are vertebrates? _____

Are any of the fossils mammals, as are humans? _____

Use the Geologic Timescale (Table 18.1) to associate each fossil with the particular era and period when this type of animal was most abundant. Fill in Table 18.3 according to sequence of the time from the latest (top) to earliest (bottom).

Summary of Fossil Record

In this section, you studied the geologic timescale. The geologic timescale gives powerful evidence of evolution because:

1. Fossils are _____.

2. Fossils can be arranged in a sequential manner because _____.

3. Younger fossils and not older fossils are more like _____.

4. In short, the fossil record shows that _____.

18.2 Comparative Anatomy

The comparative study of structures between groups of animals shows that some structures are **homologous,** they are similar in construction because they were inherited from a common ancestor. For example, the limbs in various vertebrates all have the same bones because a common ancestor had limbs with these bones. Therefore, homologous structures can be relied on to indicate evolutionary relationships, such as the ones depicted in Figure 18.1c. In contrast to homologous structures, the ones depicted in Figure 18.4 are **analogous** structures. The giant anteater and the spiny anteater are both mammals, but their snouts developed differently;

Giant anteater

Flying squirrel

fin
torpedo-body
Shark

Spiny anteater
a.

Sugar glider
b.

flipper
torpedo-body
Dolphin
c.

Figure 18.4 Analogous structures.
a. The snouts of a giant anteater and a spiny anteater are both adapted for feeding on ants, but they are not derived from a common ancestor.
b. Both flying squirrels and sugar gliders can glide from limb to limb, but this adaptation developed independently in the two groups.
c. Sharks have fins and dolphins have flippers that assist movement in the water, but these structures are not homologous and their structures are quite different.

they are similar in function but not in structure. Similarly, the sugar glider and the flying squirrel both have skin stretching between their front and hind legs, allowing them to glide as they leap from limb to limb. This adaptation to a particular environment developed independently in the two groups. Similarly, the fins of a shark and the flippers of a dolphin are adaptations to a similar environment and cannot be traced to a common ancestor.

Comparison of Adult Vertebrate Forelimbs

The limbs of all vertebrates are homologous structures. Homologous structures share a basic pattern, although there may be specific differences. The similarity of homologous structures is explainable by descent from a common ancestor.

Observation: Vertebrate Forelimbs

1. The central diagram in Figure 18.5 represents the forelimb bones of the ancestral vertebrate. The basic components are the humerus (h), ulna (u), radius (r), carpals (c), metacarpals (m), and phalanges (p) in the five digits.
2. *Carefully compare and label in Figure 18.5 the corresponding forelimb bones of the lizard, the bird, the bat, the cat, and the human.*
3. Fill in Table 18.4 to indicate which bones in each specimen appear to most resemble the ancestral condition and which differ most from the ancestral condition.

Figure 18.5   Vertebrate forelimbs.
All vertebrates evolved from a common ancestor, so their forelimbs share homologous structures.

Table 18.4   Comparison of Vertebrate Forelimbs

Animal	Bones That Resemble Common Ancestor	Bones That Differ from Common Ancestor
Lizard		
Bird		
Bat		
Cat		
Human		

4. Note the specific modifications that have occurred in some of the bones to meet the demands of a particular way of life. Relate the change in bone structure to mode of locomotion in two examples.

Example 1: _____

Example 2: _____

Conclusion: Vertebrate Forelimbs

Vertebrates are descended from a _____, but they are adapted to _____.

Comparison of Vertebrate Embryos

Vertebrates share homologous structures because they develop similarly. For example, as embryos, they all have a post-anal tail, somites (segmented blocks of mesoderm lying on either side of the notochord), and paired pharyngeal pouches and bordering gill arches. In aquatic animals, these pouches and arches become functional gills (Fig. 18.6). In humans, the first pair of pouches becomes the cavity of the middle ear and auditory tube, the second pair becomes the tonsils, and the third and fourth pairs become the thymus and parathyroid glands.

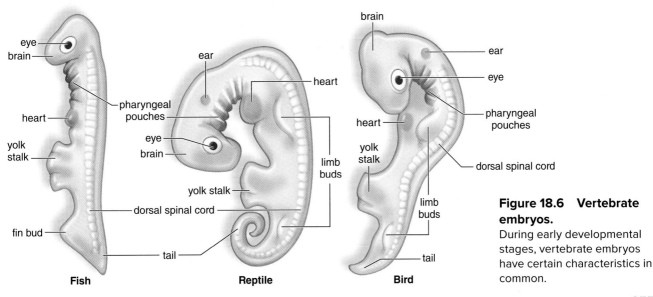

Figure 18.6   Vertebrate embryos.
During early developmental stages, vertebrate embryos have certain characteristics in common.

Observation: Chick and Pig Embryos

1. Obtain prepared slides of vertebrate embryos at comparable stages of development. Observe each of the embryos using a binocular dissecting microscope.
2. List five structural similarities of the embryos:

    a. _____

    b. _____

    c. _____

    d. _____

    e. _____

Conclusion: Vertebrate Embryos

Vertebrate embryos resemble one another because _____

_____.

Comparison of Hominid Skulls

The designation *hominid* includes humans and primates that are humanlike. Paleontologists have uncovered several fossils dated from 7.5 MYA (millions of years ago) to 200,000 years BP (before the present) when there arose humans called Cro-Magnons that are included in the hominid category (Fig. 18.7).

Figure 18.7 Hominid evolution.
Several groups of extinct hominids preceded the evolution of modern humans, designated here by color.

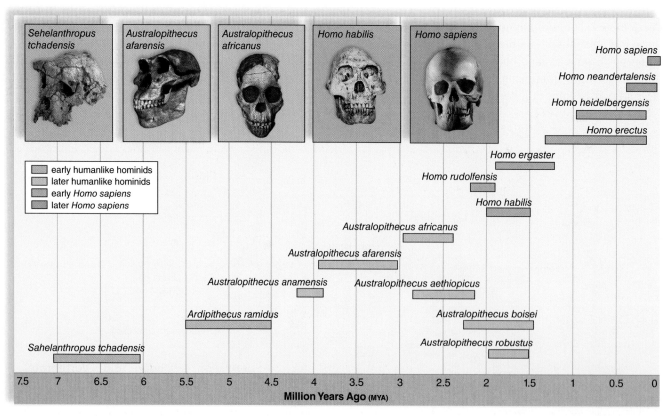

Observation of Hominid Skulls

Several of the hominid skulls noted in Figure 18.7 are on display. Use Tables 18.5, 18.6, and 18.7 to record data pertaining to the cranium (or braincase), the face, and the teeth. Compare the hominid skulls on display with the human skull (see Fig. 18.9) For example, is the forehead like or more flat than the human skull? List at least three skulls you examined for each feature listed in Tables 18.5, 18.6, and 18.7.

Table 18.5 Cranium Compared to Human Cranium			
Feature	**Skulls** **1.**	**2.**	**3.**
a. Frontal bone (like or more flat?)			
b. Supraorbital ridge (divided or continuous ridge present?)			
c. Sagittal crest (present?)			
d. Mastoid process (flat or projecting?)			

Table 18.6 Face Compared to Human Face			
Feature	**Skulls** **1.**	**2.**	**3.**
a. Nasal bones (raised or flat?)			
b. Nasal opening (larger?)			
c. Chin (projecting forward?)*			
d. Width of face (wider?)**			

*If your instructor directs you to, measure from the edge of the foramen magnum to between the incisors.
**If your instructor directs you to, measure the width of the face from mid-zygomatic arch to the other mid-arch.

Table 18.7 Dentition Compared to Human Dentition			
Feature	**Skulls** **1.**	**2.**	**3.**
a. Teeth rows (parallel or diverge from each other?)			
b. Incisors (vertical or angled?)			
c. Canine teeth (overlap other teeth?)			
d. Molars (more massive?)			

Conclusions

- Do your data appear to be consistent with the evolutionary sequence of the hominids in Figure 18.7? _____

- Report here any data you collected that would indicate this hominid was closer in time to humans than indicated in Figure 18.7? _____

- Report here any data you collected that would indicate this hominid was more distant in time from humans than indicated in Figure 18.7? _____

Comparison of Chimpanzee and Human Skeletons

Chimpanzees and humans are closely related, as is apparent from an examination of their skeletons. However, they are adapted to different ways of life. Chimpanzees are primates adapted to living in trees and are herbivores—they eat mainly plants. Humans are primates adapted to walking on the ground and are omnivores—they eat both plants and meat.

Observation: Chimpanzee and Human Skeletons

Posture

Chimpanzees are arboreal and climb in trees. While on the ground, they tend to knuckle-walk, with their hands bent. Humans are terrestrial and walk erect. Compare the posture of a chimpanzee and human (Fig. 18.8) by answering the following questions:

1. **Head and trunk:** Where are the head and trunk with relation to the hips and legs—thrust forward over the hips and legs or balanced over the hips and legs? Record your observation in Table 18.8.
2. **Spine:** Which animal has a long and S-curved lumbar region, and which has a short and slightly curved lumbar region? Record your observation in Table 18.8. How does this contribute to an erect posture in humans?

   _____

   _____

   _____

3. **Pelvis:** Chimpanzees sway when they walk because lifting one leg throws them off balance. Which animal has a narrow and long pelvis, and which has a broad and short pelvis? Record your observation in Table 18.8.
4. **Femur:** In humans, the femur better supports the trunk. In which animal is the femur angled (see arrows) to place the lower limbs under the trunk? Record your observation in Table 18.8.
5. **Knee joint:** In humans, the knee joint is modified to support the body's weight. In which animal is the femur larger at the bottom and the tibia larger at the top? Record your observation in Table 18.8.

Figure 18.8 Chimpanzee and human skeletons.
Differences in posture can be related to their adaptations to a different habitat. Chimpanzees primarily live in trees while humans walk on land.

6. Foot: In humans, the foot is adapted for walking long distances and running with less chance of injury. In which animal is the big toe opposable? How does an opposable toe assist chimpanzees?

_____

Which foot has an arch? How does an arch assist humans? _____

_____

Record your observations in Table 18.8.

7. Foramen magnum: Is the foramen magnum (opening for spine) centered in the skull or placed to the rear

of the skull? _____

Record your observations in Table 18.8. How does the difference in position of the foramen magnum

correlate with the posture and stance of the two animals? _____

_____

Table 18.8 Comparison of Chimpanzee and Human Postures		
Skeletal Part	**Chimpanzee**	**Human**
1. Head and trunk		
2. Spine		
3. Pelvis		
4. Femur		
5. Knee joint		
6. Foot: Opposable toe		
Arch		
7. Foramen magnum		

Conclusion: Chimpanzee and Human Posture

- Do your observations show that chimpanzees and humans have postures suited to different environments?
  _____ Explain. _____
  _____

Skull Features

Humans are omnivorous. A diet rich in meat does not require strong grinding teeth or well-developed facial muscles. Chimpanzees are herbivores, and a vegetarian diet requires strong teeth and strong facial muscles that attach to bony projections. Compare the skulls of the chimpanzee and the human in Figure 18.9, and answer the following questions:

Figure 18.9 Chimpanzee and human skulls.

Differences in facial features can be related to a difference in diet. Chimpanzees are herbivores and humans are omnivores.

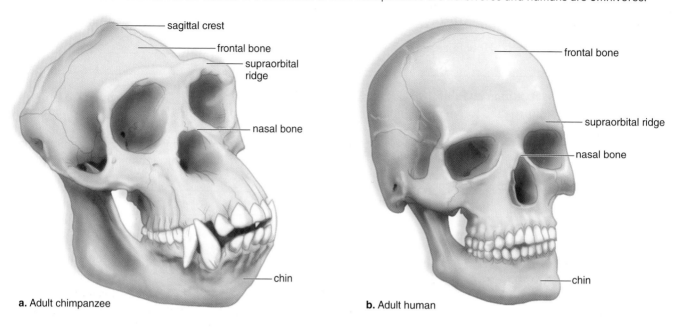

a. Adult chimpanzee

b. Adult human

1. **Supraorbital ridge:** For which skull is the supraorbital ridge (the region of frontal bone just above the eye socket) thicker? Record your observations in Table 18.9.
2. **Sagittal crest:** Which skull has a sagittal crest, a projection for muscle attachments that runs along the top of the skull? Record your observations in Table 18.9.
3. **Frontal bone:** Compare the slope of the frontal bones of the chimpanzee and human skulls. How are they different? Record your observations in Table 18.9.
4. **Teeth:** Examine the teeth of the adult chimpanzee and adult human skulls. Are the incisors (front two teeth) vertical or angled? Do the canines (adjacent to incisors) overlap the other teeth? Record your observations in Table 18.9. Are the molars larger or moderate in size? Record your observations in Table 18.9.
5. **Chin:** What is the position of the mouth and chin in relation to the profile for each skull? Record your observations in Table 18.9.

Table 18.9 Skull Features of Chimpanzees and Humans

Feature	Chimpanzee	Human
1. Supraorbital ridge		
2. Sagittal crest		
3. Shape of frontal bone		
4. Teeth		
5. Chin		

Conclusion: Skull Features

- Do your observations show that diet can be related to the facial features of chimpanzees and humans? _____ Explain. _____

Summary of Evidence from Comparative Anatomy

1. The similarity between the bones in all vertebrate forelimbs and a common ancestor shows that today's vertebrates are (descended, not descended) _____ from the common ancestor. However, the various vertebrates are adapted to _____ ways of life.

2. The similarity in the appearance of vertebrate embryos also shows that today's vertebrates are (related, not related) _____.

3. Differences in hominid skull features suggest the sequence in which these hominids _____.

4. A study of the chimpanzee skeleton and the human skeleton shows that their differences are simply due to _____ posture of humans. A contrast in skulls shows that the facial differences are due to a difference in _____.

18.3 Molecular Data

Using high-powered laboratory equipment, geneticists today can quickly compare the DNA base sequence in genes or the amino acid sequence in proteins to determine the degree to which any two groups of organisms are related. The sequence of amino acids in cytochrome c, a carrier of electrons in the electron transport chain found in mitochondria, has been determined for a variety of organisms. On the basis of the number of amino acid differences reported in Figure 18.10, it is concluded that the evolutionary relationship between humans and these groups of organisms decreases in the following order: monkeys, pigs, ducks, turtles, fishes, moths, and yeast.

This conclusion agrees with the fossil record, because the dates when these organisms are first found in the fossil record goes from yeast to monkeys (see Table 18.1). Molecular data among primates is of extreme interest because it can help determine which of the primates we are most closely related to.

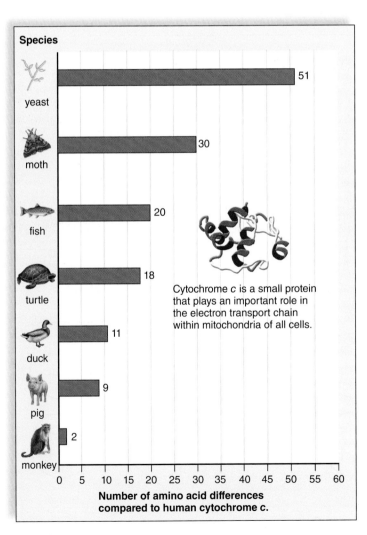

Cytochrome c is a small protein that plays an important role in the electron transport chain within mitochondria of all cells.

Number of amino acid differences compared to human cytochrome c.

Figure 18.10 Amino acid differences in cytochrome c.
The few differences found in cytochrome c between monkeys and humans show that of these organisms humans are most closely related to monkeys.

Protein Similarities

The immune system makes **antibodies** (proteins) that react with foreign proteins, termed **antigens.** Antigen–antibody reactions are specific. An antibody will react only with its particular antigen. In today's Laboratory, this reaction can be observed when a precipitate, a substance separated from the solution, appears.

Experimental Procedure: Protein Similarities

In this procedure, it is assumed that human serum (containing human antigens) is injected into the bloodstream of a rabbit, and the rabbit makes antibodies against human antigens. This experimental procedure tests this sensitized rabbit serum against the antigens of other animals. *The stronger the reaction (determined by the amount of precipitate), the more closely related the animal is to humans.*

1. Obtain a chemplate (a clear glass tray with wells), one bottle of synthetic human blood serum, one bottle of synthetic rabbit blood serum, and five bottles (I–V) of blood serum test solution.
2. Put 2 drops of synthetic rabbit blood serum in each of the six wells in the chemplate. Label the wells in the chemplate 1–6. See yellow circles in Figure 18.11.
3. Add 2 drops of synthetic human blood serum to each well. See red circles in Figure 18.11. Stir with the plastic stirring rod attached to the chemplate. The rabbit serum has now been "sensitized" to human serum. (This simulates the production of antibodies in the rabbit's bloodstream in response to the human blood proteins.) The rabbit serum now contains antibodies against human antigens.
4. Rinse the stirrer. (The large cavity of the chemplate may be filled with water to facilitate rinsing.)
5. Add 4 drops of test serum III (contains human antigens) to well 6. See blue circles in Figure 18.11.

   Describe what you see. _____

   _____

   This well will serve as the basis by which to compare all the other samples of test blood serum.

Figure 18.11 Molecular evidence of evolution.
The greater the amount of precipitate, the more closely related an animal is to humans.

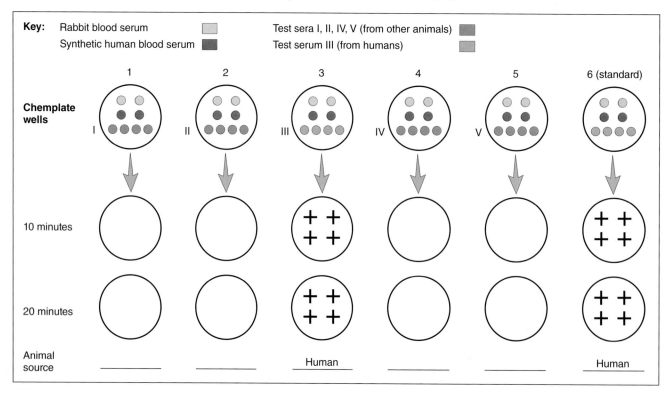

6. Now add four drops of test serum solution I to well 1. Stir and observe. Rinse the stirrer. Do the same for each of the remaining test serum solutions (II–V)—adding II to well 2, III to well 3, and so on. Be sure to rinse the stirrer after each use.

7. At the end of 10 and 20 minutes, record the amount of precipitate in each of the six wells in Figure 18.11. Well 6 is recorded as having ++++ amount of precipitate after both 10 and 20 minutes. Compare the other wells with this well (+ = trace amount; 0 = none). Holding the plate slightly above your head at arm's length and looking at the underside toward an overhead light source will allow you to more clearly determine the amount of precipitate.

Conclusions: Protein Similarities

- The last row in Figure 18.11 tells you that the test serum in well 3 is from a human. How do your test

  results confirm this? _____

- Aside from humans, the test sera (supposedly) came from a pig, a monkey, an orangutan, and a chimpanzee. Which is most closely related to humans—the pig or the chimpanzee?

  _____

- Judging by the amount of precipitate, complete the last row in Figure 18.11 by indicating which serum you believe came from which animal. On what do you base your conclusions?

  _____

Summary: Molecular Data

In Section 18.2, a comparison of bones in vertebrate forelimbs showed that vertebrates share a common ancestor. This molecular data goes one step farther and shows us that of the vertebrates studied, _____ and

_____ are most closely related. In evolution, closely related organisms share a recent common

ancestor. Humans share a more recent common ancestor with chimpanzees than they do with pigs.

_____ 1. State the simplest definition of _evolution_.

_____ 2. What does the geologic timescale encompass—tens of millions or hundreds of millions of years?

_____ 3. What is the best explanation for the fact that fossils do not resemble their modern-day representatives completely?

_____ 4. Fossils are the _____ of past life.

_____ 5. All vertebrates go through similar embryological stages. What does this suggest?

_____ 6. Which skeleton—chimpanzee or human—has a narrow and long pelvis?

_____ 7. Which has thicker supraorbital ridges, the chimpanzee skull or the human skull?

_____ 8. Which term—homologous or analogous—means that components are similar in structure?

_____ 9. During development, humans have _____, even though fish have only gill slits as adults.

_____ 10. In this laboratory, what type of molecular reaction was used to determine relatedness?

_____ 11. What does it indicate if antibodies to the serum of one species react strongly against the serum of another species?

_____ 12. All apes and humans belong to what group of mammals?

_____ 13. All vertebrates, including humans, have what type of supporting rod during embryological development?

Thought Questions

14. If a characteristic is found in bacteria, fungi, pine trees, snakes, and humans, in which organism did it most likely evolve?

15. What do mutations have to do with amino acid changes in a protein? How do such changes help determine relatedness of organisms?

16. Homologous structures arise from common ancestry; analogous structures do not. Given this information, how might you explain the existence of analogous structures?

McGraw-Hill Access Science Website

An online encyclopedia of science and technology that provides information, including videos, that can enhance the laboratory experience.

www.accessscience.com

The following LearnSmart Lab contains exercises that are related to the content of this chapter:

• Evidences of Evolution

|BIOLOGY

Find even more resources to learn the chapter concepts including animations, tutorial videos, and interactive practice questions.

LABORATORY

19

Effects of Pollution on Ecosystems

Learning Outcomes

19.1 Studying the Effects of Pollutants
- Using a hay infusion culture as the experimental material, predict the effect of
  - oxygen deprivation on species composition and diversity of ecosystems.
  - acid deposition on species composition and diversity of ecosystems.
  - enrichment on species composition and diversity of ecosystems.
- Using data from a seed germination experiment, predict the effect of acid rain on crop yield.
- Using *Gammarus* as the subject, predict the effect of acid rain on food chains involving animals.

Prelab Question: How is thermal pollution related to oxidation deprivation?

19.2 Studying the Effects of Cultural Eutrophication
- Identify potential sources of pollution and predict the effect of cultural eutrophication on the food chains of a lake or pond in your vicinity.

Prelab Question: Why do fish die off as a result of cultural eutrophication?

Application for Daily Living: **Climate Change**

Introduction

Pollution is an environmental change brought about by humans that adversely affects the health of humans and all other organisms. This laboratory will consider three causes of aquatic pollution: thermal pollution, acid pollution, and cultural eutrophication. Acid deposition (acid rain or snow) occurs after sulfur dioxide and nitrogen oxide enter the atmosphere, usually from the burning of fossil fuels. These gases are converted to acids, which return to Earth. Acid deposition kills small aquatic invertebrates and also decomposers so that the entire ecosystem is threatened. Thermal pollution occurs when water temperature rises above normal. Deforestation, soil erosion, and the burning of fossil fuels contribute to thermal pollution. The chief cause is use of water from a lake or the ocean as a coolant for the excess heat of a power plant. As water temperature rises, the amount of dissolved oxygen decreases, and most cells require a goodly supply of oxygen.

Figure 19.1 Cultural eutrophication.
Eutrophic lakes tend to have large populations of algae and rooted plants.

Cultural eutrophication is another type of pollution caused by humans. It occurs when excess chemicals enter bodies of water from agricultural fields, wastewater from sewage treatment plants, or even excess detergents. These sources of excess nutrients cause an algal bloom seen as a green scum on a lake (Fig. 19.1). When aquatic populations overgrow and die, decomposition robs the lake of oxygen, the fish die off, and the water is unfit for human consumption.

19–1

Laboratory 19   Effects of Pollution on Ecosystems   **265**

19.1 Studying the Effects of Pollutants

We are going to study the effects of pollution by observing its effects on hay infusion microorganisms (Fig. 19.2), on seed germination, and on an animal called *Gammarus.*

Study of Hay Infusion Cultures

A hay infusion culture (hay soaked in water) contains various microscopic organisms, but we will be concentrating on how the pollutants in our study affect the protozoan populations in the culture. We will consider both of these aspects: **species composition,** the number of different types of protozoans; and **species diversity,** diversity increases as the relative abundance of each type protozoan becomes equal.

Experimental Procedure: Effect of Pollutants on a Hay Infusion Culture

During this experimental procedure you will microscopically examine samples of these hay infusion cultures: (1) no treatment, i.e., the control culture; (2) oxygen deprived, i.e., heating causes water to lose oxygen; (3) acidic culture, i.e., acid has been added; and (4) enriched with nutrients, i.e., inorganic nutrients have been added. In each instance you will:

Prepare a wet mount of the culture and report the results of your examination in Table 19.1 in terms of species composition and species diversity.

1. **Control culture:** This culture simulates the species composition and diversity of a nonpolluted environment. Prepare a wet mount and with the assistance of Figure 19.2 and any guides available in the laboratory, identify as many different types of protozoans as possible in the hay infusion culture. State whether species composition is high, medium, or low. Record your estimation in the second column of Table 19.1. Do you judge species diversity to be high, medium, or low? Record your estimation in the third column of Table 19.1.

Figure 19.2   Microorganisms in hay infusion cultures.
Organisms are not to size.

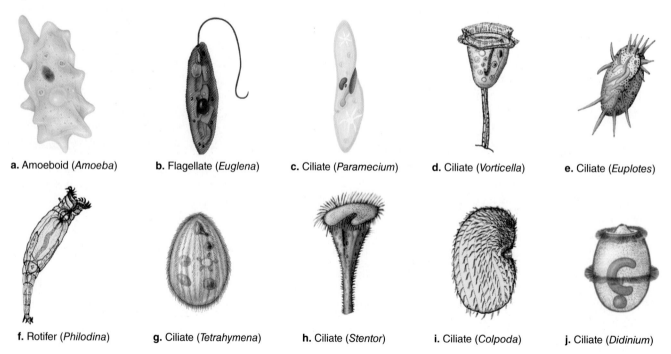

a. Amoeboid (*Amoeba*)   b. Flagellate (*Euglena*)   c. Ciliate (*Paramecium*)   d. Ciliate (*Vorticella*)   e. Ciliate (*Euplotes*)

f. Rotifer (*Philodina*)   g. Ciliate (*Tetrahymena*)   h. Ciliate (*Stentor*)   i. Ciliate (*Colpoda*)   j. Ciliate (*Didinium*)

Table 19.1 Effect of Pollution on a Hay Infusion Culture

Type of Culture	Species Composition (High, Medium, or Low)	Species Diversity (High, Medium, or Low)	Explanation
Control			
Oxygen-deprived			
Acidic			
Enriched			

2. **Oxygen-deprived culture** due to thermal pollution. In order for humans to generate electricity, power plants remove cool water from, and return warm water to, a nearby lake or ocean. When we study the effects of low oxygen on a hay infusion culture, we are studying an effect of thermal pollution, because warm water holds less oxygen than cold water.

   Prepare a wet mount of this culture and determine if there is a change in species composition and diversity. Again record the species composition and species diversity as high, medium, or low in Table 19.1.

Air and water pollution

3. **Acidic culture.** When humans burn fuels to heat their houses, or to run any sort of vehicle, or to generate electricity, acids are a by-product that enter the atmosphere, where they cause acid rain. In this culture the pH has been adjusted to 4 with sulfuric acid (H_2SO_4). This simulates the effect of acid rain on a hay infusion culture.

   Prepare a wet mount of this culture and determine if there is a change in species composition and diversity. Again record the species composition and species diversity as high, medium, or low in Table 19.1.

4. **Enriched culture.** Organic nutrients enter water from sewage treatment plants—even detergents contain inorganic nutrients— and from agricultural fields and our lawns due to the use of fertilizers. These nutrients will cause the algae population, which is food for most protozoans, to increase. In the short term, their species composition should increase. Eventually, as the algae die off, decomposition will rob the water of oxygen and the protozoans may start to die off.

Water pollution

   Prepare a wet mount of this culture and determine if there is a change in species composition and diversity. Again record the species composition and species diversity as high, medium, or low in Table 19.1.

Conclusions

- What could be a physiological reason for the adverse effects of oxygen deprivation on a hay infusion culture? If consistent with your results, enter this explanation in the last column of Table 19.1.
- What could be a physiological reason for the adverse effects of a low pH on a hay infusion culture? If consistent with your results, enter this explanation in the last column of Table 19.1.
- What could be a physiological reason for the adverse effects of an enriched culture? If consistent with your results, enter this explanation in the last column of Table 19.1.

Figure 19.3 Acid rain reduces crop yield.

Effect of Acid Rain on Seed Germination

Seeds depend on favorable environmental conditions of temperature, light, and moisture to germinate, grow, and reproduce. Like any other biological process, gernimation requires enzymatic reactions that can be adversely affected by an unfavorable pH. Acid rain adversely affects the health of plants, animals, and humans (Fig. 19.3). For example, many forests are now dying, some lakes are devoid of fish, and humans possibly suffer more bronchial disorders (e.g., asthma) due to acid rain.

Experimental Procedure: Effect of Acid Rain on Seed Germination

In this experimental procedure, we will test whether there is a negative correlation between acidic water and germination. In other words, it is hypothesized that as acidity increases, the more likely seeds

will _____.

Your instructor has placed 20 sunflower seeds in each of five containers with water of increasing acidity: 0% vinegar (tap water), 1% vinegar, 5% vinegar, 20% vinegar, and 100% vinegar.

1. Test and record the pH of solutions having the vinegar concentrations noted above. Record the pH of each solution in Table 19.2.
2. Count the number of germinated sunflower seeds in each container, and complete Table 19.2.

Table 19.2 Effect of Acidity on Germination

Vinegar	pH	Percent Germination
0%		
1%		
5%		
20%		
100%		

Conclusions: Effect of Acidity on Germination of Sunflower Seeds

- As you know, each enzyme has an optimum pH. Explain why acid rain is expected to inhibit metabolism, and therefore, seedling development. _____

_____

- Do the data support or falsify your hypothesis? _____

Study of *Gammarus*

A small crustacean called *Gammarus* lives in ponds and streams (Fig. 19.4) where it feeds on debris, algae, or anything smaller than itself, such as some of the protozoans in Figure 19.2. In turn, fish like to feed on *Gammarus.*

Experimental Procedure: Gammarus

Add 25 ml of spring water to a beaker and record the pH of the water. _____ pH

Add four *Gammarus* to the container. Do they all use their legs in swimming? _____

Which legs are used in jumping and climbing? _____

- What do *Gammarus* do when they "bump" into each other? _____
_____

Figure 19.4   *Gammarus.*

Gammarus Control Sample

After observing *Gammarus,* decide what two behaviors are most often observed. During a 5-minute time span, total the amount of time spent doing each of these behaviors.

Behaviors	Amount of Time	Total Time
1._____	_____	_____
2._____	_____	_____

Gammarus Test Sample

If so directed by your instructor, put a *Gammarus* in a beaker of spring water adjusted to pH 4 by adding vinegar. During a 5-minute time span, total the amount of time spent doing each of the same behaviors.

Behaviors	Amount of Time	Total Time
1._____	_____	_____
2._____	_____	_____

Conclusion: Gammarus

- Draw a conclusion from this study. _____
_____
_____

- Create a food chain that shows who eats whom and includes algae, protozoans, *Gammarus,* fish, and humans. _____

   a. Predict what would happen to this food chain if the water was

      oxygen deprived. _____

      acidic. _____

      enriched with inorganic nutrients (short term and long term). _____
_____

Conclusions: Studying the Effects of Pollutants

- Give an example to show that the hay infusion study pertains to real ecosystems. _____
  _____

- What are the potential consequences of acid rain on crops that reproduce by seeds? _____
  _____

  On the food chains of the ocean? _____

- How does the addition of nutrients affect species composition and species diversity of an ecosystem over
  time? _____

19.2 Studying the Effects of Cultural Eutrophication

Chlorella, the green alga used in this study, is considered to be representative of algae in bodies of freshwater. The protozoan *Daphnia* feeds on green algae such as *Chlorella*. First, you will observe how *Daphnia* feeds, and then you will determine the extent to which *Daphnia* could keep the effects of cultural eutrophication from occurring in a hypothetical example. Keep in mind that this case study is an oversimplification of a generally complex problem.

Observation: Daphnia *Feeding*

1. Place a small pool of petroleum jelly in the center of a small petri dish.
2. Use a dropper to take a *Daphnia* from the stock culture, place it on its back (covered by water) in the petroleum jelly, and observe it under the stereomicroscope.
3. Note the clamlike carapace and the legs waving rapidly as the *Daphnia* filters the water.
4. Add a drop of carmine solution, and observe how the *Daphnia* filters the "food" from the water and passes it through the gut. The gut is more visible if you push the animal onto its side. In this position, you may also observe the heart beating in the region above the gut and just behind the head.
5. Allow the *Daphnia* to filter-feed for up to 30 minutes, and observe the progress of the carmine particles through the gut. Does the carmine travel completely through the gut in 30 minutes?

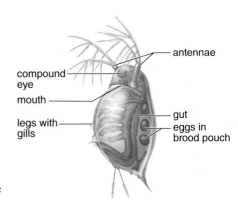

Experimental Procedure: Daphnia *Feeding on* Chlorella

This exercise requires the use of a spectrophotometer. Absorbance will be a measure of the algal population level; the greater the number of algal cells, the greater the absorbance. The higher the absorbance, the greater the amount of light absorbed and *not* passed through the solution.

1. Obtain two spectrophotometer tubes (cuvettes) and a Pasteur pipette.
2. Fill one of the cuvettes with distilled water, and use it to zero the spectrophotometer. Save this tube for number 6.
3. Use the Pasteur pipette to fill the second cuvette with *Chlorella*. Gently aspirate and expel the sample several times (without creating bubbles) to give a uniform dispersion of the algae.
4. Add ten hungry *Daphnia,* and following your instructor's directions, immediately measure the absorbance with the spectrophotometer. If a *Daphnia* swims through the beam of light, a strong deflection should

occur; do not use any such higher readings—instead, use the lower reading for the absorbance. Record your reading here. _____

5. Remove the cuvette with the *Daphnia* to a safe place in a test-tube rack. Allow the *Daphnia* to feed for 30 minutes.

6. Rezero the spectrophotometer with the distilled water cuvette.

7. Measure the absorbance of the experimental cuvette again. Record your reading here and explain.

_____

_____.

Experimental Procedure: Case Study in Cultural Eutrophication

The following problem will test your understanding of the value of a single species—in this case, *Daphnia.* Please realize that this is an oversimplification of a generally complex problem.

1. Assume that developers want to build condominium units on the shores of Silver Lake. Homeowners in the area have asked the regional council to determine how many units can be built without causing the lake to become eutrophic. As a member of the council, you have been given the following information.

    The present population of *Daphnia,* 10 animals/liter, filters 24% of the lake per day, meaning that it removes this percentage of the algal population per day. This is sufficient to keep the lake essentially clear. Predation—the eating of the algae—will allow the *Daphnia* population to increase to no more than 50 animals/liter. What percentage of the lake per day would 50 Daphnia/liter be able to clear?

    _____

2. The sewage system of the condominiums will add nutrients to the lake. Phosphorus (P) output will be 1 kg per day for every ten condominiums. This will cause a 30% increase in the algal population. Using this information, complete Table 19.3.

Table 19.3 Cultural Eutrophication		
Condominiums	(P) Added	Increase in Algae
10	1 kg	30%
20		
30		
40		
50		

Conclusion: Studying the Effects of Cultural Entrophication

- Assume that phosphorus (P) is the only nutrient that will cause an increase in the algal population and that *Daphnia* is the only type of zooplankton available to feed on the algae. How many condominiums would you allow the developer to build? _____

- What other possible impacts could condominium construction have on the condition of the lake?

_____

Application for Daily Living

Climate Change

Climate change includes an expected rise in temperature of the Earth's surface and oceans due to the burning of fossil fuels, such as oil and gasoline. When we burn fossil fuels, carbon dioxide enters the atmosphere and traps heat from the sun's rays near the Earth.

Any significant rise in the Earth's temperature can lead to all sorts of disastrous results (Fig. 19.5). The sea level can rise and flood cities from Miami to New York on the East Coast and San Diego and San Francisco on the West Coast. Rain patterns can change, and some areas that

Agricultural land parched by lack of rain

Houses threatened by rising water

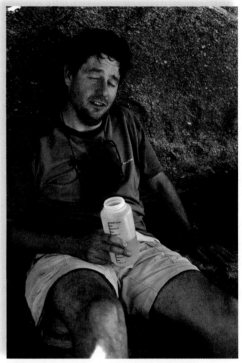
Human fatigued by exertion in hot weather

Figure 19.5 Some effects of climate change.

were dry before can become wet and wet areas can become dry. The favorable temperature for farming can move northward to regions that don't have good soil; plants and animals that can't migrate will become extinct. The soil in Alaska is called a permafrost because it is always frozen. If the glaciers and permafrost melt, animals such as the polar bear are threatened, and so is the Alaskan pipeline that helps supply American cities.

Even if you don't believe climate change is imminent, you should do all you can to reduce consumption of oil. When we burn less oil, air pollution lessens, we are not as dependent on less-than-friendly foreign countries, and it may even help reduce the national debt.

———————— 1. Any effect on seeds would typify an effect on what type organism in an ecosystem?

———————— 2. What type pollution exercise negatively affected both plants and animals equally?

———————— 3. Why did you observe *Gammarus* in spring water before acidic water?

———————— 4. What type of pollution results when water from rivers and ponds is used to cool industrial processes?

———————— 5. In your experiment, did you add acid or base to adjust the hay infusion culture to pH 4?

———————— 6. What condition does acid deposition cause that can be harmful to organisms?

———————— 7. What would a negative control in a cultural eutrophication study lack?

———————— 8. Overenrichment causes which types of populations to increase in size beyond the ordinary?

———————— 9. In the case study, what two factors caused the producer population (*Chlorella*) to increase?

———————— 10. What type instrument was used to relatively measure the amount of phytoplankton in your test tube?

———————— 11. What final effect might excess nutrients have on aquatic ecosystems?

Thought Questions

12. Does pollution affect all living things, including humans? How?

13. When pollutants enter the environment, they have far-ranging effects. Give an example from this laboratory.

14. Why might it be important to monitor predator/prey interactions when studying ecosystems?

Find even more resources to learn the chapter concepts including animations, tutorial videos, and interactive practice questions.

McGraw-Hill Access Science Website

An online encyclopedia of science and technology that provides information, including videos, that can enhance the laboratory experience.

www.accessscience.com

The following LearnSmart Lab contains exercises that are related to the content of this chapter:

• Effects of Pollution on Ecosystems

METRIC SYSTEM

Unit and Symbol	Metric Equivalent	English Equivalents	Units of Temperature

Length

nanometer (nm)	$= 10^{-9}$ m $(10^{-3}\ \mu m)$		
micrometer (μm)	$= 10^{-6}$ m $(10^{-3}$ mm)		
millimeter (mm)	$= 0.001\ (10^{-3})$ m		
centimeter (cm)	$= 0.01\ (10^{-2})$ m	1 inch $= 2.54$ cm	
		1 foot $= 30.5$ cm	
meter (m)	$= 100\ (10^{2})$ cm	1 foot $= 0.30$ m	
	$= 1{,}000$ mm	1 yard $= 0.91$ m	
kilometer (km)	$= 1{,}000\ (10^{3})$ m	1 mile $= 1.6$ km	

Weight (mass)

nanogram (ng)	$= 10^{-9}$ g		
microgram (μg)	$= 10^{-6}$ g		
milligram (mg)	$= 10^{-3}$ g		
gram (g)	$= 1{,}000$ mg	1 ounce $= 28.3$ g	
		1 pound $= 454$ g	
kilogram (kg)	$= 1{,}000\ (10^{3})$ g	$= 0.45$ kg	
metric ton (t)	$= 1{,}000$ kg	1 ton $= 0.91$ t	

Volume

microliter (μl)	$= 10^{-6}$ l $(10^{-3}$ ml)		
milliliter (ml)	$= 10^{-3}$ l	1 tsp $= 5$ ml	
	$= 1$ cm^3 (cc)	1 fl oz. $= 30$ ml	
	$= 1{,}000$ mm³		
liter (l)	$= 1{,}000$ ml	1 pint $= 0.47$ l	
		1 quart $= 0.95$ l	
		1 gallon $= 3.79$ l	
kiloliter (kl)	$= 1{,}000$ l		

Units of Temperature scale:

°F			°C
	230	110	
	220		
212° —	210	100 — 100°	
	200		
	190	90	
	180	80	
	170		
160° —	160	70 — 71°	
	150		
	140	60	
134° —	130	57°	
	120	50	
105.8° —	110	40 — 41°	
98.6° —	100	37°	
	90	30	
	80		
	70	20	
56.66° —	60	13.7°	
	50	10	
	40		
32° —	30	0 — 0°	
	20		
	10	−10	
	0	−20	
	−10		
	−20	−30	
	−30		
	−40	−40	

Common Temperatures*

°C	°F	
100	212	Water boils at standard temperature and pressure.
57	134	Highest recorded temperature in the United States, Death Valley, July 10, 1913
41	105.8	Average body temperature of a marathon runner in hot weather
37	98.6	Human body temperature
0	32.0	Water freezes at standard temperature and pressure.

* See temperature scale

To convert temperature scales:
$$°C = (°F - 32)/1.8$$
$$°F = 1.8(°C) + 32$$

HINTS FOR USING THE METRIC SYSTEM

1. Use decimals, not fractions (e.g., 2.5 m, not 2½ m). Express measurements in units requiring only a few decimal places. For example, 0.3 m is more easily manipulated and understood than 300000000 nm.
2. When measuring pure water, the metric system offers an easy and common conversion from volume measured in liters to volume measured in cubic meters to mass measured in grams: 1 ml = 1 cm^3 = 1 g.
3. Do not place a period after metric symbols (e.g., 1 g, not 1 g.). Use a period after a symbol only at the end of a sentence. Do not mix units or symbols (e.g., 9.2 m, not 9 m 200 mm). Metric symbols are always singular (e.g., 10 km, not 10 kms).
4. Except for degrees Celsius, always leave a space between a number and a metric symbol (e.g., 20 mm, not 20mm; 10°C, not 10° C).
5. Use a zero before a decimal point when the number is less than one (e.g., 0.42 m, not .42 m).

Photo Credits

Laboratory 1

Figure 1.1: © Ryan McVay/Getty RF;
1.3: © Denkou Images/age fotostock RF;
1.4: © Carolyn A. McKeone/Science Source.

Laboratory 2

Figure 2.1: © Hill Street Studios/Harmik
Nazarian/Getty RF; 2.2: © Bruce M. Joèson;
2.5: © Martyn F. Chillmaid/Science Source;
2.6(both): © Richard Hutchings; 2.8: © Arno
Massee/Science Source.

Laboratory 3

Figure 3.2a: © Michael Ross/Science Source;
3.2b: © CNRI/SPL/Science Source; 3.2c:
© Steve Gschmeissner/Science Source;
3.3–3.4: © Leica Microsystems GmbH;
3.5(both): © Ricochet Creative Productions
LLC; 3.8: © Dr. Gopal Murti/Science Source;
3.9: © M.I.(Spike) Walker/Alamy; 3.10 a, c:
© Martin Rotker/Science Source; 3.10b: © SPL/
Science Source.

Laboratory 4

Page 35(food): © Joè Thoeming/McGraw Hill
Education; p. 36(tea cup): © George Doyle/
Getty RF; 4.2a: © Don W. Fawcett/Science
Source; p. 37(iodine test): © Martin Shields/
Science Source; 4.3: © Jeremy Burgess/SPL/
Science Source; p. 39(Benedict's test): © David
S. Moyer; p. 40(hemoglobin): © P. Motta
& S. Correr/Science Source; p. 40(hair):
© Goldmund Lukic/Getty RF; p. 42(biuret test):
© David S. Moyer; 4.7–4.9: © Joè Thoeming/
McGraw-Hill Education.

Laboratory 5

Figure 5.5(all): © David M. Phillips/Science
Source.

Laboratory 6

Figure 6.1: © Ed Rescçe; 6.2: © Garry
DeLong/Science Source; 6.3–6.6 and p. 65a–d:
© Ed Rescçe; 6.7: © Dennis Strete/McGraw-
Hill Education; 6.8–6.9(bone): © Ed Rescçe/
Getty Images; 6.9(loose, hyaline, dense,
adipose)–6.10: © Ed Rescçe; 6.11–6.12: © Al
Telser/McGraw-Hill Education; 6.13: © Dennis
Strete/McGraw-Hill Education; 6.14a: © Ed
Rescçe; 6.16: © imago stock&people/Newscom.

Laboratory 7

Figure 7.6: © Jim Womack/McGraw-Hill
Education; 7.7: © Ken Taylor/Wildlife Images.

Laboratory 8

Page 91(heart, lungs dissection): © Professor
David F. Cox/McGraw-Hill Education; 8.1–8.2:
© Eric Wise; 8.3b: © Ed Rescçe; 8.4: © Ilene
MacDonald/Alamy RF; 8.5a–b, d–e: © Ed
Rescçe; 8.5c: © Biophoto Associates/Science
Source; 8.6a: © Ed Rescçe; 8.6c: © Biophoto
Associates/Science Source.

Laboratory 9

Figure 9.2b: © Ed Rescçe.

Laboratory 10

Page 129(man): © AP Photo/Ivan Chavez.

Laboratory 11

Figure 11.5a: © Dwight Kuè; 11.8: © Phipps
& Bird, Inc., Richmond, VA.

Laboratory 12

Figure 12.2(hyaline, bone): © Ed Rescçe;
12.2(osteocyte): © Biophoto Associates/Science
Source; 12.5: © Corbis RF; 12.7b: © Eric
Wise; p. 162(skeletal muscle): © Ed Rescçe;
12.12(gymnast): © Corbis RF; 12.12(sarcomere):
© Biology Media/Science Source.

Laboratory 13

Figure 13.2(all): © Dr. J. Timothy Cannon;
13.5: © Kage-mikrofotografie/Phototake;
13.7: © P.H. Gerbier/SPL/Science Source.

Laboratory 14

Page 183(zygote): © Anatomical Travelogue/
Science Source; p. 183(embryo, 1 week):
© Bettman/Corbis; p. 183(embryo, 8 weeks):
© Neil Harding/Getty Images; p. 183(fetus,
3 months): © Petit Format/Science Source;
p. 183(fetus, 5 months): © Joè Watney/Science
Source; 14.2b–14.4: © Ed Rescçe; 14.5a:
© Dr. Robert Calentine/Visuals Unlimited/
Corbis; 14.5b–f and 14.9: © Carolina Biological
Supply Company/Phototake; 14.12: © Martin
Rotker/Phototake; p. 197(both): © Petit format/
Science Source.

Laboratory 15

Figure 15.2a: © Andrew Syred/Science Source;
15.4(all): © Ed Rescçe; 15.5(cleavage):
© National Institutes of Health(NIH)/USHHS;
15.5(contractile): © Steve Gschmeissner/SPL/
Getty RF.

Laboratory 16

Figure 16.2a: © Superstock; 16.2b: © HFPA,
63rd Golden Globe Awards; 16.2c–f: © Bob
Coyle/McGraw-Hill Education; 16.2g: © Corbis
RF; 16.2h: © Creatas/PunchStock RF;
p. 225(Down syndrome): © Stockbyte/Veer RF;
16.1: © CNRI/SPL/Science Source; p. 226(Poly
X): © The McElligott Family; p. 226(Turner):
© Tarheel Images/UNC Medical Illustration and
Photography; p. 226(Klinefelter): © Stefan D.
Schwarz, http://klinefeltersyndrome.org;
p. 226(Jacob): © The Giles Family; 16.10(all):
© CNRI/SPL/Science Source.

Laboratory 17

Figure 17.10(both): © Bill Longcore/Science
Source.

Laboratory 18

Table 18.1, p. 249(woolly mammoths):
© Gianni Dagli Orti/Corbis; Table 18.1,
p. 250(frog with insect): © Dwight Kuè;
18.2(ammonite): © Carl Pendle/Getty Images;
18.2(echinoderms): © DEA/G. Nimatallah/
Getty Images; 18.2(mantis,pseudoscorpion):
© Joè Cancalosi/Getty Images; 18.2(snails):
© Ed Rescçe/Getty Images; 18.2(trilobites):
© Danita Delimont/Getty Images; 18.3(bird):
© WaterFrame/Alamy; 18.3(fish, snake):
© Joè Cancalosi/Getty Images; 18.3(deerlike
mammal): © Gary Ombler/Getty Images;
18.3(frog): © DEA/G. Cigolini/Getty Images;
18.3(duckbill dinosaur): © Kevin Schafer/
Getty Images; 18.4a(giant anteater): © Tom
Brakefield/Corbis; 18.4a(spiny anteater): © Tom
McHugh/Science Source; 18.4b(flying squirrel):
© Joe McDonald/Corbis; 18.4b(sugar glider):
© ANT Photo Library/Science Source;
18.4c(shark): © Brand X Pictures/PunchStock
RF; 18.4c(dolphin): © Jeffrey L. Rotman/Getty
Images; 18.5(man): © Eric Wise/McGraw-Hill
Education; 18.5(bird): © McGraw-Hill
Education; 18.5(bat): © Jack Milchanowski/
Getty Images; 18.5(cat): © Marc Henrie/Getty
Images; 18.5(lizard): © Mauricio Handler/Getty
Images; 18.7(S. tchadensis): © HO/AFP/
Newscom; 18.7(A. afarensis): © Friedrich
Saurer/Alamy; 18.7(A. africanus): © Philippe
Plailly/Science Source; 18.7(H. habilis): © Kike
Calvo VWPics/Superstock; 18.7(H. sapiens):
© Kenneth Garrett/Getty Images.

Laboratory 19

Figure 19.1: © Pat Watson/McGraw-Hill
Education; p. 267(air pollution): © Mike
Theiss/Getty Images; p. 267(water pollution):
© fotog/Getty RF; 19.3: © Tony Bee/Getty
Images; 19.4: © William Amos/Photoshot;
19.5(drought): © AP Photo/Seth Periman;
19.5(rising water): © AP Photo/HO/Mike Page,
www.mike-page.co.uk; 19.5(heat exhaustion):
© TS Corrigan/Alamy.

Index

Note: Page references followed by *f* and *t* refer to figures and tables, respectively.

A

Abdomen skin fold, 126
Abdominal aorta, 102*f*
Abdominal cavity, fetal pig, 81, 81*f*, 83, 85–88
Abduction, 160, 160*f*
ABO blood typing, 98–99, 224–225
Absolute dating, 251
Accessory organs, 106*f*
Accommodation, 176
Acetabulum, 158*f*
Achilles tendon, 159*f*
Acid deposition, 265, 267
Acidic cultures, 266, 267
Acidic groups, 40
Acromion process, 157*f*
Acrosomes, 185, 185*f*
Actin, 50*f*, 69, 162, 163, 163*f*
Active sites, 8, 58*f*
Active transport, 53, 53*f*
Activities, energy cost, 120*t*
Adduction, 160, 160*f*
Adductor longus, 159*f*
Adenine, 233
Adipose tissue, 63*f*, 67, 68*f*
Afferent arterioles, 134*f*, 135, 135*t*
Agglutination, 99
Agranular leukocytes, 98
Air, speed of diffusion in, 52
Albumin, 40, 42, 107
Alcohol consumption, 147, 214
Algae, eutrophication and, 265, 265*f*, 267, 270–272
Algae blooms, 265, 265*f*
Allantois, 195, 196*f*
Alleles, 217, 224–225
Alveoli, 138, 139*f*, 148
Alzheimer disease, 225
Amino acids
    abbreviations, 239*t*
    absorption, 112
    evolution evidence from, 261, 261*f*
    harmful base sequence changes, 243–244, 243*f*
    sequences in proteins, 236, 238–239
    structures, 40–41, 41*f*
    in tRNA, 238–239, 238*f*
Amino groups, 40, 143–144
Ammonite fossils, 251*f*
Amnion, 191*f*, 193, 193*f*, 195
Amniotic cavity, 191*f*, 195, 196*f*
Amniotic fluid, 195
Amoeba, 266*f*
Amoebozoans, 266*f*
Amphibians, 252*f*
Amylase, 40
Analogous structures, 253, 253*f*
Anaphase, 203*f*, 204, 210*f*
Anaphase I, 205, 206*f*, 210*f*
Anaphase II, 206*f*, 209
Animal cells, 50–51
Animal evolution, 248*f*
Ankles, fetal pig, 78*f*, 81*f*

Antagonistic pairs, 150, 161
Anteaters, 253, 253*f*
Antennae, *Dapāia*, 270*f*
Anterior, defined, 77
Anterior compartment, 174*f*
Antibodies, 40, 98, 262
Antibody-antigen reactions, 262
Anticodons, 238
Antigens, 98, 224–225, 262
Anus
    in female, 186*f*
    fetal pig, 78*f*, 79, 81*f*, 85, 86, 87
    location, 76*f*, 106*f*
    in male, 184*f*
Aorta
    fetal pig, 88*f*
    human, 92, 92*f*, 93*f*
    in systemic circuit, 101*f*
Aortic arch, 92, 92*f*
Aortic semilunar valve, 93, 93*f*
Apex of heart, 92*f*
Appendicular skeleton, 152, 152*f*, 157–158, 157*f*–158*f*
Appendix, 76*f*, 77
Aquaporins, 55
Aqueous humor, 174*f*, 174*t*
Archenteron, 190*f*
Arm bones, 157
Arm of microscope, 24
Arteries
    anatomy, 99*f*
    coronary, 92, 92*f*
    fetal pig, 80
    functions, 99
    liver, 143, 143*f*, 144, 145
    major, 102*f*
    pulmonary and systemic, 101*f*
Arterioles, 99, 100, 132, 132*f*
Arteriovenosus shunt, 99, 99*f*
Arthropods, 251*f*, 270, 270*f*
Aspartic acid, 239*t*
Asters, 200, 201*f*, 201*t*, 202*f*
Atria, 92, 92*f*, 93*f*
Atrioventricular valves, 93
Attached earlobes, 217*f*, 218*f*, 219*t*
Auditory association area, 171*f*
Auditory canal, 177*f*, 177*t*
Auditory tube, 177*f*, 177*t*, 255
Australopithecus afarensis, 256*f*
Australopithecus africanus, 256*f*
Autosomes
    commonly inherited traits, 218–219, 218*f*, 219*t*, 221*f*
    defined, 218
    inherited disorders, 221–222
Average daily energy intake, 116–118
Average daily energy needs, 119–123
AV node, 94, 94*f*
Axial skeleton, 152, 153–156, 153*f*–156*f*
Axilla-side skin fold, 126
Axons, 71, 71*f*
Axon terminals, 166*f*

B

Balance scales, 13–14, 14*f*
Basal metabolic rate, 121, 122, 123
Basal metabolism, 115, 121
Basement membrane, 62*f*, 64*f*, 65*f*
Base pairing, 232, 232*f*, 234
Basophils, 97*f*, 98
Beaumont, William, 105
Benedict's reagent, 39–40, 39*t*, 55, 111, 144
Bicarbonate ions, 139
Biceps brachii, 149–150, 149*f*, 159, 159*f*
Biceps femoris, 159*f*
Bicuspid valve, 93, 93*f*
Bile, 85, 109, 109*f*, 147
Bile canals, 143*f*
Bile duct, 85, 86, 109*f*, 143*f*
Bile salts, 44
Binocular head of microscope, 21, 22, 24*f*
Biotecèology, 231
Birds, 252, 252*f*, 255*f*
Biuret reagent, 41, 107
Blastocoel, 188, 190*f*
Blastocyst, 189, 189*f*
Blastomere, 202
Blastopore, 190*f*
Blastula, 188, 188*f*, 190*f*, 202
Blind spot, 175
Blood
    cell types, 69, 69*f*, 97–98, 97*f*
    as connective tissue, 69, 69*f*
    glucose regulation, 144–146, 147
    microscopic observations, 29–30
    pH, 139
    tonicity, 56–57, 59
    in urine, 138
Blood disorders, 223, 243–245
Blood pressure, 95–97, 103, 132
Blood type, 98–99, 224–225
Blood vessels. *See also* Arteries; Veins
    anatomy, 99–100, 99*f*
    basic functions, 91
    fetal pig, 80
    systems, 100–101, 101*f*
    umbilical, 195
Body composition, 126–129
Body mass index, 125
Body surface area, 121*f*, 122
Bone marrow, 150*f*, 151, 165
Bone marrow transplants, 165
Bones
    appendicular skeleton, 152, 152*f*, 157–158, 157*f*–158*f*
    axial skeleton, 152, 152*f*, 153–156, 153*f*–156*f*
    homologous, 254–255, 254*f*
    long bone anatomy, 150–151, 150*f*
    measuring, 10–11, 11*f*
Bone tissue
    basic features, 67, 67*f*, 68*f*, 150*f*
    in long bones, 150–151

Brachiocephalic artery, 92*f*
Brain
    in central nervous system, 167, 167*f*
    embryonic, 192, 255*f*
    human brain anatomy, 170–171, 171*f*
    sheep brain dissection, 168–170, 169*f*
Brain stem, 169*f*, 170
Breads, 117*t*
Breathing, 140–142
Brittle stars, 251*f*
Bronchi, 77, 83
Bronchioles, 138
Bulbourethral gland, 88, 184, 184*f*
Burger King, 118*t*
Burn patients, 73

C

Calories, 47, 116, 117*t*, 118*t*
Cambrian period, 250*t*
Canadian Home Fitness Test, 6*t*
Canaliculi, 67, 67*f*, 68*f*, 150*f*, 151
Cancer, 32, 165, 214
Canines, 79*f*
Capillaries
    fetal, 195
    functions, 99
    observing, 100
    in portal systems, 101
    pulmonary, 138, 139*f*
    systemic, 140*f*
Capillary beds, 99, 99*f*
Capillary exchange, 140*f*
Capitulum, 157*f*
Carbohydrates
    in foods, 35, 36, 46*f*
    structure, 36*f*
    testing for, 37–40
Carboniferous period, 249*t*
Cardiac muscle, 70, 70*f*
Cardiac stress tests, 7
Cardiac veins, 93
Cardiovascular disease, 43, 130
Cardiovascular system
    basic functions, 91
    coordination with other systems, 89
    heart, 3–7, 92–95
    organs, 76*f*
Carotid arteries, 101*f*, 103
Carpals, 152*f*, 157, 157*f*
Carriers, 53, 221, 222
Cartilage, 68, 68*f*, 150*f*, 151
Catalase, 58–59
Caudal fold of amnion, 193, 193*f*
Cecum, 86
Cell bodies of neurons, 71, 71*f*
Cell cycle, 200–204, 200*f*
Cells
    comparative sizes, 11*f*
    diffusion in, 52–57
    overview, 49
    structures, 50–51, 50*f*

Cellular metabolism, 40
Cellular stages of development, 188–189, 188*f*, 189*f*
Celsius scale, 16, 16*f*
Cenozoic era, 249*t*, 250
Centimeter, 10, 10*t*, 12
Central canal
    bone, 67, 67*f*, 68*f*, 150*f*, 151
    spinal cord, 172*f*, 173*f*
Central nervous system
    brain anatomy, 168–171, 169*f*, 171*f*
    components, 167, 167*f*
    development, 192
Central sulcus, 168, 169*f*, 171*f*
Central tendon, fetal pig, 81
Centrioles, 50*f*, 200, 201*f*, 201*t*
Centromeres, 199, 200, 201*f*, 201*t*, 203
Centrosomes, 50*f*, 201*f*, 201*t*
Cerebellum, 168, 169*f*
Cerebral hemispheres, 168, 169*f*
Cerebrospinal fluid, 168
Cerebrum, 168, 171*f*
Cervical curvature, 155*f*
Cervical vertebrae, 155, 155*f*
Cervix, 32, 186*f*
Chambers, heart, 92
Cheek cells, 31, 31*f*, 49
Chemical exposure, 214
Chemoreceptors, 139, 180
Chemplates, 262, 262*f*
Chemstrip test strips, 137–138
Chest skin fold, 126
Chick embryos, 192–193, 193*f*
Chicken Ringer solution, 192
Chimpanzees, 258–260, 258*f*, 260*f*
Chin, human versus chimpanzee, 260, 260*f*
Chlorella, 270–271
Cholesterol, 130
Chondrocytes, 68, 150*f*
Chordae tendineae, 93, 93*f*
Chorion, 195, 196*f*
Chorionic villi, 195, 196*f*
Choroid, 174*f*, 174*t*
Chromatids, 201*t*, 235
Chromatin, 50*f*
Chromosomes
    autosomal, 218
    defined, 201*t*
    duplication, 199, 200, 201*f*
    reduction in meiosis, 199, 205
    viewing anomalies, 212, 225–226, 225*f*, 226*f*
Cilia, 62, 65, 65*f*
Ciliary body, 174*f*, 174*t*
Ciliates, 266*f*
Circulatory system. *See* Cardiovascular system
Circumduction, 160*f*
Cirrhosis, 147
Clavicles, 152*f*, 156*f*, 157, 157*f*
Cleavage, 188, 188*f*, 189*f*
Cleavage furrows, 203*f*, 204, 204*f*
Climate change, 272
Coal-forming forests, 249*t*
Coarse-adjustment knob, 25, 26
Coccyx, 152*f*, 155, 155*f*
Cochlea, 177, 177*f*, 177*t*
Cochlear nerve, 177*f*
Codons, 238
Cold receptors, 178*f*, 179

Collagen, 66, 66*f*, 68*f*, 100
Collecting ducts, 133*f*, 134, 134*f*
Colon, fetal pig, 85, 86
Color blindness, 222–223, 223*f*
Colpoda, 266*f*
Common bile duct, 109*f*
Common carotid arteries, 82*f*, 84*f*, 102*f*
Common hepatic duct, 109*f*
Common iliac artery, 102*f*
Common iliac vein, 102*f*
Compact bone
    basic features, 67, 67*f*, 68*f*
    in long bones, 150*f*, 151
Comparative anatomy, 253–261
Complementary base pairing, 232, 232*f*, 234
Compound eyes, 270*f*
Compound light microscope, 19, 20, 21*t*, 23–29
Concentration, effect on enzyme activity, 58
Conclusions, 3
Condensers (microscope), 25
Cone cells, 174*t*, 175
Connective tissue, 66–69, 66*f*–69*f*, 72
Contractile ring, 204
Contractions, muscular, 161–164
Controlled variables, 4
Controls, 2, 36, 106, 266
Coracoid process, 157*f*
Cornea, 174*f*, 174*t*, 181
Coronary arteries, 92*f*, 93
Corpus callosum, 168, 169*f*, 171*f*
Corpus luteum, 187, 187*f*
Costal cartilages, 152*f*, 156*f*
Coxal bones, 152*f*, 158, 158*f*
Cranial nerves, 167, 167*f*, 169*f*, 172
Cranium, 153, 154, 257*t*
Crenation, 56, 56*f*
Cretaceous period, 249*t*
Cro-Magnons, 256
Cross-bridges, myosin, 163, 163*f*
Crossing-over, 205
Cross sections, 19, 19*f*
Crustaceans, 269
Cultural eutrophication, 265, 270–272
Curvatures of spine, 155*f*
Cystic fibrosis, 221
Cytochrome *c*, 261, 261*f*
Cytokinesis, 199, 200, 204, 204*f*
Cytoplasm
    defined, 49
    division during mitosis, 199, 204, 204*f*
    in generalized cell, 50*f*
Cytosine, 233
Cytoskeletons, 50*f*

D

Dapāia, 270–272, 270*f*
Data, defined, 2
Daughter cells, 199, 200, 207*f*
Daughter chromosomes, 200, 203*f*
Degradation reactions, 58, 58*f*
Dehydration, 59
Deletions, chromosome, 212
Deltoid muscle, 159, 159*f*
Deltoid tuberosity, 157*f*
Dendrites, 71, 71*f*

Dense fibrous connective tissue, 66, 66*f*, 68*f*
Deoxyribonucleic acid (DNA)
    isolating, 241–242
    protein synthesis, 236–240, 239*f*, 240*f*
    replication, 199, 232, 234–235
    RNA versus, 236*t*
    structure, 231, 232–233, 232*f*
    triplet code, 238
Dependent variables, 4
Depth of field (microscope), 28–29, 29*f*
Dermis, 63, 63*f*
Devonian period, 250*t*
Diabetes mellitus, 138, 144
Diagnoses, 32
Diameter of field (microscope), 27–28
Diaphragm
    fetal pig, 80, 81*f*, 84*f*
    location, 76*f*
    microscope, 25, 26
    role in inspiration, 140, 141*f*
Diastole, 95
Diastolic pressure, 103
Didinium, 266*f*
Diencephalon, 168, 169*f*
Diet
    nutrition labeling, 49
    weight control and, 129–130
Differential blood cell count, 32
Diffusion, 52–55
Digestion
    fats, 43, 43*f*, 109–110
    proteins, 107–108
    requirements for, 113
    starch, 36, 36*f*, 111–112
Digestion reactions, 36*f*, 41*f*
Digestive system
    coordination with other systems, 89
    fetal pig, 85
    organs, 76–77, 76*f*, 106*f*
    tissues, 72, 72*f*
Digital scales, 13, 13*f*
Digits, fetal pig, 78*f*, 81*f*
Dinosaurs, 249*t*, 252*f*
Dipeptides, 41*f*
Diploid cells, 205, 212
Disaccharides, 36*f*, 39
Discrimination, touch, 179
Dissecting microscopes, 20, 21–23, 22*f*
Dissection instructions, fetal pig, 77, 80–81, 81*f*
Distal convoluted tubules, 134, 134*f*
DNA. *See* Deoxyribonucleic acid (DNA)
DNA sequencing, 243, 244, 245
Dolphins, 253*f*, 254
Dominant alleles, 217
Dominant disorders, 221, 222
Dorsal, defined, 77
Dorsal horn, 173*f*
Dorsal root ganglia, 173*f*
Dorsal spinal cord, 255*f*
Double eyelashes, 228–229
Double helix, 232*f*, 233
Down syndrome, 225
Duckbill dinosaur, 252*f*
Duodenum, 85, 113

Duplications, chromosome, 212

E

Earlobes, 177*f*, 217*f*, 218*f*, 219*t*
Early gastrula stage, 190
Ears, 78*f*, 81*f*, 176–178
ECGs, 7
Echinoderms, 251*f*
Ecosystems, polluted, 265–272
Ectoderm, 190, 190*f*, 191*f*
Efferent arterioles, 134*f*, 135, 135*t*
Eggs
    Dapāia, 270*f*
    development, 187, 187*f*
    fusion with sperm, 189*f*, 212, 213*f*
Ejaculatory ducts, 184, 184*f*
Elastic fibers, 66*f*, 68*f*, 100
Elbows, fetal pig, 78*f*, 81*f*
Electrocardiograms, 94–95, 94*f*
Electrode placement, 95
Electron microscopes, 19, 20–21, 20*f*, 49
Embryonic disk, 191, 196*f*
Embryos
    development, 183, 188–194, 193*f*, 194*f*
    environmental influences, 194
    similarities among vertebrates, 255–256, 255*f*
Emphysema, 148
Emulsification, 44–45, 44*f*, 109, 109*f*
Emulsions, 44, 45*f*
Endocrine system, 76*f*, 132, 132*f*
Endoderm, 190, 190*f*, 191*f*
Endoplasmic reticulum, 50*f*, 51
Endothelium, 100
Energy
    average daily intake, 116–118
    average daily requirements, 119–123
    balancing intake with need, 123
Environmental influences on embryos, 194
Enzymatic proteins, 40
Enzymes
    basic functions, 40, 58
    in cells, 49
    disabling, 105
    environmental effects on, 58–59
    protein digestion, 107–108
Eocene epoch, 249*t*
Eosinophils, 97*f*, 98
Epidermis, 62, 63*f*
Epididymides, 87, 88*f*, 184, 184*f*, 185*f*
Epiglottis, 76*f*, 79*f*, 80
Epithelial tissue
    major functions, 62, 72
    microscopic observations, 31, 31*f*
    types, 62–65, 62*f*–65*f*
Epochs, 250
Eras, 250
Erectile tissue, 184*f*
Erythrocytes
    basic features, 69, 69*f*, 97
    blood typing, 98–99
    crenation and hemolysis, 56–57, 56*f*
    disorders, 223, 243–245

Erythrocytes—*(Cont.)*
    in lung tissue, 138
    microscopic observations, 20*f*,
        29, 30
    in urine, 138
Esophagus
    fetal pig, 80, 80*f*, 82, 83, 85
    functions, 77, 106*f*
    location, 76*f*
Ethmoid bone, 153*f*
Euglena, 31, 31*f*, 266*f*
Euplotes, 266*f*
Eversion, 160, 160*f*
Evolution
    defined, 247
    evidence from comparative
        anatomy, 253–261
    fossil evidence, 248–253
    molecular evidence, 261–263
    plants, 249*t*–250*t*
Exercise, 5–7
Experimental variables, 4–5
Experiments, 2, 36
Expiration, 140, 141*f*
Expiratory reserve volume, 141, 142
Extension, 160, 160*f*
Extensor carpi group, 159*f*
Extensor digitorum, 159, 159*f*
Extensor digitorum longus, 159*f*
External auditory canal, 153*f*
External genitals, 79, 197
External jugular vein, 102*f*
External oblique, 159*f*
Extinctions, 249*t*, 250*t*
Extraembryonic membranes,
    195, 196*f*
Eyelashes, 228–229
Eyepiece lenses, 21, 23
Eyes
    accommodation, 176
    anatomy, 174–176, 174*f*, 174*t*
    blind spot, 175
    embryonic, 255*f*
    fetal pig, 78*f*, 81*f*
Eyespots, *Euglena*, 31*f*

F

Facial bones, 154, 257*t*
Facilitated transport, 53, 53*f*
Factor VIII, 223
Fahrenheit scale, 16, 16*f*
Fallopian tubes, 86
False ribs, 156*f*
Fast-food restaurants, 118*t*
Fats
    digestion, 43, 43*f*, 109–110
    energy content, 117*t*
    in foods, 47
    as percentage of body weight,
        126–129
    testing for, 44
Fatty acids, 43, 43*f*, 112
Female reproductive system
    fetal pig, 86–87, 87*f*
    human, 186–187, 186*f*
Femoral artery, 102*f*
Femoral vein, 102*f*
Femur, 150*f*, 152*f*, 158, 158*f*,
    258, 258*f*
Fertilization, 189, 189*f*, 209,
    212, 213*f*
Fertilization membrane, 188*f*

Fetal pig
    dissection instructions, 77,
        80–81, 81*f*, 89
    external anatomy, 77–79, 78*f*
    neck region, 82–83, 82*f*
    oral cavity and pharynx, 79–80,
        79*f*, 80*f*
    reproductive systems, 86–88,
        87*f*, 88*f*
    respiratory system, 83, 84*f*
    thoracic and abdominal cavities,
        80–81, 81*f*, 83–88, 84*f*
    transplants from, 75
    urinary system, 86, 87*f*
Fetus, human, 196*f*, 197, 197*f*, 197*t*
Fiber in foods, 47
Fibroblasts, 66*f*, 68*f*
Fibula, 152*f*, 158, 158*f*
Field of view, 27
Filtrate, glomerular, 135
Fimbriae, 86, 186*f*, 189*f*
Fine-adjustment knob, 25, 26
Finger length, 218*f*
Finger straightness, 219*t*
Fishes, 252*f*, 255*f*
Fixed eyepiece, 21
Flagella, *Euglena*, 31*f*
Flagellates, 266*f*
Flexion, 160, 160*f*
Flexor carpi group, 159*f*
Floating ribs, 156, 156*f*
Flowering plants, 249*t*
Fluid replacement, 59
Flying squirrel, 253*f*, 254
Focusing knobs, 21, 22, 25
Fogle, Jared, 129
Follicles, 186, 187, 187*f*
Food and Drug Administration
    (U.S.), 35
Food chains, 269
Food diaries, 115, 116
Foods
    energy content, 117*t*, 118*t*
    nutrition labeling, 47
    safety, 35
    testing, 46
    weight control and, 130
Foot, human versus chimpanzee,
    258*f*, 259
Foramen magnum, 153, 153*f*, 259
Forearm bones, 157
Forebrain, 193*f*
Forelimbs
    fetal pig, 78*f*, 81*f*
    homologous structures,
        254–255, 254*f*
Foreskin, 184*f*
Forests, 249*t*
Fossil fuels, 272
Fossils, 248–253
Fourth ventricle, 169*f*, 171*f*
Fovea centralis, 174*f*, 174*t*
Freckles, 218*f*, 219*t*
Free nerve endings, 178*f*
Frogs, 190, 190*f*, 252*f*
Frontal bone
    human versus chimpanzee,
        260, 260*f*
    location, 152*f*, 153, 153*f*,
        154, 154*f*
Frontalis, 159

Frontal lobe, 168, 169*f*, 171*f*
Fruits, energy content, 117*t*

G

G₁ stage, 200, 200*f*
G₂ stage, 200, 200*f*
Gallbladder
    fetal pig, 85, 86
    functions, 77, 106*f*, 109*f*
    location, 76*f*
Gametes, 212–214, 213*f*
Gametogenesis, 199, 212–214, 213*f*
Gammarus, 269, 269*f*
Gas exchange in lungs, 139,
    139*f*, 140*f*
Gastric glands, 107*f*
Gastric pits, 107*f*
Gastrocnemius, 159*f*
Gastrointestinal tract, 72, 72*f*
Gastrula, 190, 190*f*
Gel electrophoresis, 241–242,
    242*f*, 244*f*
Gender selection, 229
Genetic counseling, 225–229
Genetic disorders
    detecting, 243–245
    genetic counseling and, 225–229
    inheritance, 220–225
Genetics
    evidence for evolution, 261–263,
        262*f*
    genetic counseling, 225–229
    genotype determination,
        218–219
    inheritance, 220–225
    overview, 217
Genitals, 79, 197
Genomes, 243
Genomic sequencing, 243, 244, 245
Genotype, 217, 218–219
Geologic timescale, 248, 249*t*–250*t*,
    250–251
Germ cells, 185, 185*f*
Germination, 268
Germ layers, 183, 191, 191*f*
Giant anteater, 253, 253*f*
Gill arches, 255
Glans penis, 184*f*
Glenoid cavity, 157*f*
Glomerular capsule, 133, 134*f*
Glomerular filtration, 133, 134*f*, 135
Glomerulus, 134*f*, 135, 135*t*
Glottis, 76*f*, 77, 80, 80*f*
Glucagon, 144
Glucose
    molecular structure, 36*f*
    in polysaccharide formation, 36
    reaction to Benedict's reagent,
        39*t*
    regulation by liver, 144–146,
        147
    in urine, 138
Glutamic acid, 239*t*, 244*f*
Gluteus maximus, 159, 159*f*
Glycerol, 43*f*, 163
Glycogen, 36, 37*f*, 132, 147
Goblet cells, 64, 64*f*, 65*f*
Golgi apparatus, 50*f*
Gonads, 187. *See also* Ovaries;
    Testes
Graafian follicles, 187
Graduated pipets, 15, 15*f*

Gram, 13
Granular leukocytes, 98
Gray matter, 172, 172*f*, 173*f*
Greater trochanter, 158*f*
Greater tubercle, 157*f*
Great saphenous vein, 102*f*
Green perception, 223
Growth plates, 150*f*
Guanine, 233
Gut, *Dapāia*, 270, 270*f*

H

Hair cells, 177
Hair follicles, 63*f*
Hairline, 218*f*, 219*t*
Haploid cells, 205, 212
Hard palate, 76*f*, 79, 79*f*
Hay infusion cultures, 266–268
Health. *See also* Genetic disorders
    blood pressure and, 103
    body systems, 89
    cancer risks, 214
    cardiovascular, 7
    hydration for, 59
    liver's importance, 147
    nutrition labeling, 49
    weight control and, 129–130
Health care
    bone marrow transplantation,
        165
    engineered tissues, 73
    eye surgery, 181
    microscopic observations in, 32
    stress tests, 7
Hearing, in humans, 177–178
Heart
    anatomy, 92–94, 92*f*, 93*f*
    blood and blood flow, 97–102
    Dapāia, 270
    embryonic development,
        192–193, 193*f*
    fetal pig, 83, 84*f*
    location, 76*f*, 77
    physiology, 94–95, 94*f*
    pulse and blood pressure, 95–97
Heart attacks, 7
Heartbeat, 94, 94*f*, 95–97
Heart rate experiment, 3–7
Heart sounds, 95
Heat receptors, 178*f*, 179
Height and weight tables, 124–125
Hemispheres, cerebral, 168, 169*f*
Hemoglobin, 40, 243–245
Hemolysis, 56, 56*f*, 57*f*
Hemophilia, 223
Hepatic artery, 143, 143*f*
Hepatic cells, 143, 143*f*
Hepatic portal vein, 101*f*, 143, 143*f*,
    144, 145
Hepatic veins, 101*f*, 144, 145
Hepatitis, 147
Herbivores, 259
Heterozygous individuals, 217, 217*f*
High blood pressure, 103
High-power diameter of field, 28
High-power objective, 25
Hindbrain, 193*f*
Hindlimbs, fetal pig, 78*f*, 81*f*
Hitchhiker's thumb, 219*t*
Holocene epoch, 249*t*
Homeostasis
    kidneys' role, 133–138

liver's role, 143–146
negative feedbacks in, 132, 132*f*
overview, 131, 131*f*, 146
respiration and, 138–142
Hominids, 249*t*, 256–260, 256*f*
Homogenization, 44
Homo habilis, 256*f*
Homologous chromosomes
(homologues), 205,
206*f*–207*f*
Homologous structures, 253,
254–255, 254*f*
Homo sapiens, 256*f*
Homozygous dominant individuals,
217, 217*f*
Homozygous recessive individuals,
217, 217*f*
Human Genome Project, 243
Humans
chimpanzees versus, 258–260,
258*f*, 260*f*
evolution, 256–257, 256*f*
first appearance, 249*t*, 256
Humerus, 149*f*, 152*f*, 156*f*,
157, 157*f*
Huntington disease, 221–222
Hyaline cartilage, 68, 68*f*, 150*f*, 151
Hydrogen bonds, 233
Hydrogen peroxide, 58
Hypertension, 103
Hypertonic solutions, 55, 56*f*
Hypothalamus, 168, 169*f*
Hypotheses, 2–3, 4
Hypotonic solutions, 55, 56*f*

I

Ideal weight, 124–129
Identical twins, 189
Iliac arteries, 101*f*
Iliac veins, 101*f*
Ilium, 158, 158*f*
Ilium-hip skin fold, 126
Illuminators (microscope), 22
Implantation, 189*f*
Incisions, 80–81
Incisors, 79*f*
Incus, 177, 177*f*
Independent focusing eyepiece,
21, 23
Inferior mesenteric artery, 102*f*
Inferior vena cava
functions, 92, 101*f*
location, 92*f*, 102*f*
relation to liver, 143*f*
Inguinal canal, 87, 88*f*
Inheritance
common human traits,
218*f*, 219*t*
determining, 220–225
genetic counseling and, 225–229
Inner cell mass, 189, 189*f*
Inner ear, 177*f*, 177*t*
Inner layer, blood vessel, 99*f*, 100
Insertions, muscle, 149, 149*f*
Inspiration, 140, 141*f*
Inspiratory reserve volume,
141, 142
Insulin, 40, 132, 132*f*, 138, 144
Integumentary system, 62–63
Intercalated disks, 70, 70*f*
Intercostal muscles, 141*f*
Interkinesis, 207*f*, 209

Intermediate filaments, 50*f*
Internal environment, 131
Internal jugular vein, 102*f*
Interneurons, 172, 173*f*
Interphase, 199, 200
Interstitial cells, 185, 185*f*
Intervertebral disks, 155*f*
Intervertebral foramina, 155*f*
Intestines, 84*f*, 85–86, 143*f*. See
also Large intestine; Small
intestine
Inversion in microscopes, 26
Inversion of foot, 160, 160*f*
Inversions, chromosome, 212
Invertebrates, 251, 251–252, 251*f*
Involuntary muscle, 70
Iodine test, 37–38
Iris, 174*f*, 174*t*
Ischium, 158, 158*f*
Isometric contractions, 161–162
Isotonic contractions, 161
Isotonic solutions, 55, 56*f*

J

Jacob syndrome, 226
Joint movements, 160, 160*f*
Jugular veins, 101*f*, 102*f*
Jurassic period, 249*t*

K

Karyotypes, 225, 225*f*, 226*f*
Keratin, 40, 63
Ketones in urine, 138, 144
Kidneys
epithelial tissue, 64*f*
fetal pig, 86, 87*f*, 88*f*
location, 76*f*
major functions, 77, 135–136
structure, 133–134
Kilocalories, 116, 117*t*, 118*t*, 129
Kilogram, 13
Kinetochores, 201*t*
Klinefelter syndrome, 226
Knee-jerk reflex, 173, 173*f*
Knee joints, 68*f*, 81*f*, 258, 258*f*
Krause end bulbs, 178*f*

L

Lacrimal bone, 153*f*
Lacunae
in cartilage, 68, 150*f*, 151
in compact bone, 67, 67*f*, 68*f*,
150*f*, 151
Lamellae, 67*f*, 150*f*, 151
Lanugo, 197
Large intestine
fetal pig, 84*f*, 85, 86
functions, 77, 106*f*
location, 76*f*
relation to liver, 143*f*
Larynx
fetal pig, 80*f*, 82, 82*f*, 84*f*
functions, 77
location, 76*f*
LASIK eye surgery, 181
Late gastrula stage, 190
Lateral epicondyle, 158*f*
Lateral malleolus, 158*f*
Lateral sulcus, 171*f*
Lateral ventricle, 169*f*, 171*f*
Latissimus dorsi, 159*f*
Lean body weight, 126–129

Left atrioventricular valve, 93, 93*f*
Left atrium, 92, 92*f*
Left cardiac vein, 92*f*
Left common carotid artery, 84*f*
Left pleural cavity, 83
Left pulmonary artery, 92, 92*f*
Left pulmonary veins, 92*f*
Left ventricle, 92, 92*f*, 93*f*
Legumes, 117*t*
Length measurement, 10–12
Lenses, microscope, 20, 21, 22,
24–25
Lens of eye, 174*f*, 174*t*, 176, 181
Lesser trochanter, 158*f*
Letter *e* slides, 26
Leucine, 239*t*
Leukocytes. *See* White blood cells
Life, evolution, 248*f*
Ligaments, 66, 151, 174*f*
Light microscopes, 19, 20–21
Lipids
emulsification, 44–45, 109
in foods, 35, 46*f*
structure, 43, 43*f*
testing for, 44
Liquids, 14–15, 15*f*, 52
Liter, 14
Liver
fetal pig, 84*f*, 85
importance, 147
location, 76*f*, 143*f*
main functions, 77, 106*f*, 147
role in digestion, 109, 109*f*
role in homeostasis, 132, 132*f*,
143–146
Lobes, cerebral, 171, 171*f*
Lobes of liver, 143
Lobules, liver, 143, 143*f*
Long bone anatomy, 150–151, 150*f*
Longitudinal fissure, 169*f*
Longitudinal sections, 19, 19*f*
Loose fibrous connective tissue, 66,
66*f*, 68*f*
Lower limb bones, 158
Low-power diameter of field, 28
Low-power objective, 25
Lumbar curvature, 155*f*
Lumbar vertebrae, 155, 155*f*
Lung cancer, 214
Lungs
epithelial tissue, 62*f*
fetal pig, 83, 84*f*
location, 76*f*
major functions, 139–140
respiratory volumes, 140–142
structure, 138, 138*f*
Lymphatic system, 76*f*
Lymphocytes, 20*f*, 97*f*, 98
Lysosomes, 50*f*

M

Magnesium ions, 163
Magnification, 27
Magnification changing knobs,
22, 23
Male reproductive system
fetal pig, 87–88, 88*f*
human, 184–185, 184*f*
Malleus, 177, 177*f*
Maltose, 36, 36*f*, 39*t*
Mammals, 249*t*, 252, 252*f*
Mammary glands, 79

Mandible, 152*f*, 153*f*, 154, 154*f*
Mantis fossils, 251*f*
Manubrium, 156*f*
Marrow, 150*f*, 151, 165
Masseter, 159*f*
Mass extinctions, 249*t*, 250*t*
Matrix
cartilage, 68*f*, 150*f*, 151
compact bone, 67, 150*f*, 151
Maxilla, 152*f*, 153*f*, 154, 154*f*
McDonald's, 118*t*
Meats, 117*t*
Mechanical stage of microscope,
25
Medial epicondyle, 158*f*
Medial malleolus, 158*f*
Medulla oblongata, 169*f*, 170
Medullary cavity, 150*f*, 151
Meiosis
in gametogenesis, 199
mitosis versus, 210–212,
210*f*–211*f*, 211*t*
overview, 205
phases, 205–209
Meiosis I, 205–208, 206*f*–207*f*,
210*f*–211*f*
Meiosis II, 209, 211*f*
Meissner corpuscles, 178*f*
Meninges, 171*f*
Menisci, 15, 15*f*
Merkel discs, 178*f*
Mesenteric arteries, 101*f*, 144, 145
Mesenteric veins, 102*f*
Mesenteries, 85
Mesoderm, 190, 190*f*, 191*f*
Mesozoic era, 249*t*
Messenger RNA, 231, 236, 237*f*
Metabolism, 40
Metacarpals, 152*f*, 157, 157*f*
Metaphase, 202*f*, 203–204,
203*f*, 210*f*
Metaphase I, 205, 206*f*, 210*f*
Metaphase II, 206*f*, 209
Metatarsals, 152*f*, 158, 158*f*
Meter, 10, 10*t*
Metersticks, 10–11, 27
Methionine, 239*t*
Methylene blue, 31
Metric system, 9–15, 27, 27*t*, 30
Micrometer, 10, 10*t*, 11–12, 27
Microscopes
clinical applications, 32
compound light microscope
features, 23–29, 24*f*
electron versus light, 20–21
making observations with,
29–31
overview, 19
rules for using, 25
stereomicroscope features, 20,
21–23, 22*f*
Microtubules, 50*f*
Microvilli, 62
Midbrain, 169*f*, 170, 193*f*
Middle ear, 177*f*, 177*t*, 255
Middle layer, blood vessel, 99*f*, 100
Milk, 44, 117*t*
Milligram, 13
Milliliter, 14
Millimeter, 10, 10*t*, 11–12
Miocene epoch, 249*t*
Mitochondria, 50*f*, 163*f*

...ed structures, 200–202,
   ...01t
...rined, 199
   meiosis versus, 210–212,
      210f–211f, 211t
   phases, 203–204, 203f–204f
   purposes, 199
Mitral valve, 93
Molecular evidence for evolution,
   261–263
Molecular genetics, 231, 261–263
Mollusks, 251f
Monocytes, 97f, 98
Monosaccharides, 39
Morula, 188, 188f, 189f, 190f
Motor neurons, 71f, 172, 173f
Mouth, 76, 79, 106f, 270f
M stage, 200, 200f
Mucosa, intestinal, 72, 72f
Mucus, 64, 64f, 65f
Multiple-allele traits, 224–225
Muscle fibers, 161, 162–164
Muscles
   of breathing, 141f
   contractions and movements,
      160–164
   naming, 159–160
   superficial skeletal muscles,
      159f
   tissues, 69–70, 69f–70f, 72
   types, 149
Muscularis, 72, 72f
Musculoskeletal system overview,
   149–150
Mutations, 214
Myelin sheath, 71f
Myofibrils, 162, 163f
Myosin, 40, 69, 162, 163, 163f

N

Nanometer, 10, 10t, 27
Nasal bone, 153f, 154, 154f, 260f
Nasal passages, 77
Nasopharynx, 76f, 77, 79f, 80
Near point, 176
Neck, fetal pig, 82–83
Negative controls, 36, 106
Negative feedback, 132
Negative tests, 36
Nephrons
   basic functions, 133
   circulation, 134f, 135–136
   structure, 133–134, 133f, 134f
Nervous system
   central nervous system anatomy,
      168–172, 169f, 171f
   chemoreceptors, 180
   development, 192
   ears and hearing, 176–178, 177f
   eyes, 174–176
   genetic disorders, 221–222
   negative feedback, 132, 132f
   overview, 167
   peripheral nervous system
      anatomy, 172–173, 173f
   skin receptors, 63f, 178–179
   tissues, 71, 71f, 72, 167
Neural tube, 190f, 192, 193, 193f
Neurofibromatosis, 221
Neuroglia, 71, 71f
Neuromuscular junction, 166f
Neurons, 71, 71f, 172

Neurulation, 190f
Neutral pH, 58
Neutrophils, 69, 69f, 97f, 98
Nipples, fetal pig, 78f, 79
Nonpolar molecules, 44
Nose, fetal pig, 78f, 81f
Nosepiece of microscope, 24
Notochord, 190f, 191f, 192
Nuclear envelope, 50f
Nuclei, 50f, 51, 201t
Nucleoli, 50f, 201t, 203
Nucleotides, 232, 232f
Nutrients as pollutants, 265, 267
Nutrients in foods, 117t
Nutrition labeling, 47

O

Obesity, 113, 125
Objective lenses, 20, 23, 24–25
Observations, 2
Occipital bone, 152f, 153, 153f
Occipital lobe, 168, 169f, 171f
Ocular lenses, 20, 23
Oil consumption, 272
Oil glands, 63f
Oil immersion objective, 25
Oils, digestion, 43
Olfactory bulb, 168, 169f
Oligocene epoch, 249t
Omnivores, 259
Oocytes, 212, 213f
Oogenesis, 187, 187f, 212, 213f
Optic chiasma, 168, 169f
Optic nerve, 168, 174f, 174t, 175
Oral cavity, fetal pig, 79–80, 79f
Orbicularis oculi, 159f
Orbicularis oris, 159f
Ordovician period, 250t
Organelles, 49
Organs
   defined, 61
   development, 192–194,
      193f, 194f
   muscle tissue, 70
   tissue structure, 72
Organ systems, 73, 76–77, 76f,
   89, 106f
Origins, muscle, 149, 149f
Osmosis, 55–57
Ossicles, 177f, 177t
Osteoblasts, 150f
Osteocytes, 67, 67f, 68f, 151
Osteons, 67, 67f, 68f, 150f, 151
Outer ear, 177f, 177t
Outer layer, blood vessel, 99f, 100
Oval window, 177f
Ovarian vein, 87f
Ovaries
   fetal pig, 86, 87f
   human, 186, 186f, 189f
Overweight, 125
Oviducts, 86, 186, 186f
Ovulation, 189f
Oxygen-deprived waters,
   266, 267

P

Pacemaker, 94
Pacinian corpuscles, 178f
Pain receptors, 178f
Palatine bone, 153f, 154
Paleocene epoch, 249t
Paleozoic era, 249t–250t

Pancreas
   digestive functions, 77, 85, 106f,
      109–112, 109f
   endocrine functions, 77, 85,
      132, 132f, 144
   fetal pig, 85, 86
   location, 76f, 106f
Pancreatic amylase, 111–112
Pancreatic duct, 109f
Pancreatic juice, 109f
Pancreatic lipase, 109–110
Papa Joè's, 118t
Papillary muscles, 93f
Pap test, 32
Paramecium, 266f
Parathyroid gland, 82f, 255
Parent cells, 200
Parfocal microscopes, 26
Parietal bones, 152f, 153, 153f
Parietal lobe, 168, 169f, 171f
Partial hydrogenation, 43, 130
Particulate genetics, 217
Patella, 158, 158f
Patellar tendon, 173
Paternity, determining, 224–225
Peanut butter, 117t
Pectoral girdle, 152f, 156f,
   157, 157f
Pectoralis major, 159f
Pedigrees, 227–229
Pelvic curvature, 155f
Pelvic girdle, 152f, 158, 158f
Pelvis, human versus chimpanzee,
   258, 258f
Penis
   fetal pig, 79, 87, 88, 88f
   human, 184, 184f
Pepsin, 107–108
Peptide bonds, 41, 41f
Peptides, 41
Pericardial cavity, 83
Pericardial membrane, 83
Periods, 250
Periosteum, 150f, 151
Peripheral nervous system, 167,
   167f, 172–173, 173f
Peritoneum, 85, 86
Peritubular capillary network, 134f,
   135, 135t, 136
Permafrost, 272
Permian period, 249t
Personal DNA sequencing, 245
Pesticides, 214
pH
   effect on enzyme activity, 58–59
   impact on germination, 268
   maintenance in blood, 139
   in stomach, 107
Phalanges, 152f, 157, 157f,
   158, 158f
Pharyngeal pouches, 255, 255f
Pharynx
   fetal pig, 80, 80f
   functions, 77, 106f
   location, 76f
Phenol red solution, 110, 241
Phenotype, 217
Phenylalanine, 222
Phenylketonuria, 222
Philodina, 266f
pH indicator, 110
Phospholipids, 44
Photomicrographs, 20, 20f

Photoreceptors, 175
Physical activity diaries, 115,
   119, 120t
Pineal body, 169f
Pinna, 177f, 177t
Pipets, 15, 15f
Pituitary gland, 171f
Placenta, 195, 196f
Plant evolution, 249t–250t
Plaque, 103, 130
Plasma, 69, 69f
Plasma membrane, 49, 50f, 53–55
Platelets, 69, 97f
Pleistocene epoch, 249t
Pleural membrane, 83
Pliocene epoch, 249t
Poisons, 147
Polar bodies, 212, 213f
Polar molecules, 44, 45f
Poles, spindle, 200, 201f, 202f–203f
Pollution, 265–272
Polypeptides, 41, 238
Polyribosomes, 50f
Polysaccharides, 36, 37f
Poly-X syndrome, 226
Pons, 169f, 170
Portal systems, 101
Positive controls, 36, 106
Positive tests, 36
Post-anal tail, 255
Posterior, defined, 77
Posterior compartment, 174f
Potassium ions, 163
Potassium permanganate, 52
Potato, 38, 38f, 56
Precambrian time, 250t
Precapillary sphincters, 99, 99f
Prefrontal area, 171f
Prehypertension, 103
Premotor area, 171f
Presbyopia, 176
Pressure receptors, 178f
Primary auditory area, 171f
Primary follicles, 186, 187f
Primary motor area, 171f
Primary olfactory area, 171f
Primary oocytes, 186, 187f, 213f
Primary somatosensory area, 171f
Primary taste area, 171f
Primary visual area, 171f
Primates, 248f, 249t, 256–260
Proline, 239t
Prophase, 202f, 203, 210f
Prophase I, 205, 206f, 210f
Prophase II, 206f, 209
Prostate gland, 88, 184, 184f
Proteins
   digestion, 41, 107–108, 107f
   in foods, 35, 46f
   major functions, 40
   similarities between species,
      262–263
   structure, 41f
   synthesis, 236–240, 239f, 240f
   testing for, 41–42
   in urine, 138
Protoslo, 31
Protozoans, 266–267
Proximal convoluted tubules,
   133, 134f
Pseudoscorpion, 251f
Pseudostratified ciliated columnar
   epithelium, 65, 65f

Pseudostratified epithelium, 62
Pubic bone, 184f, 186f
Pubis, 158, 158f
Pulmonary arteries and veins, 92, 92f, 101f
Pulmonary capillaries, 138, 139f, 140f
Pulmonary circuit, 100–101, 101f
Pulmonary semilunar valve, 93, 93f
Pulmonary trunk, 92, 92f
Pulse rate, 3–4, 5–6
Punnett squares, 220
Pupil, 174f, 174t
Purkinje fibers, 94, 94f
P wave, 94, 94f

Q

QRS wave, 94, 94f
Quadriceps femoris, 159f
Quaternary period, 249t
Quickening, 197

R

Rabbit serum, 262–263
Radial artery, 5
Radiation exposure, 214
Radius, 149f, 152f, 157, 157f
Receptors. See Sensory receptors
Recessive alleles, 217
Recessive disorders, 221, 222
Rectum
    fetal pig, 85, 86, 87
    functions, 106f
    human, 186f
Rectus abdominis, 159, 159f
Red blood cells. See Erythrocytes
Red marrow, 150f, 151, 165
Red perception, 223
Reflexes, spinal, 172–173
Regulatory proteins, 40
Relative dating, 251
Renal arteries, 101f, 102f, 133f, 134f
Renal cortex, 133, 133f
Renal medulla, 133, 133f
Renal pelvis, 133, 133f
Renal pyramids, 133f
Renal veins, 101f, 102f, 133f, 134f
Replication of DNA, 232, 234–235
Reproductive system
    female anatomy, 186–187, 186f
    fetal pig, 86–88, 87f, 88f
    male anatomy, 184–185, 184f
Reptiles
    birds as, 252, 252f
    embryos, 255f
    fossils, 252f
    on geologic timescale, 249t
Resolution (microscope), 21
Respiratory system
    coordination with other systems, 89
    fetal pig, 83, 84f
    organs, 76f, 77
    respiratory volumes, 140–142
    role in homeostasis, 138–142
Respiratory volumes, 140–142
Retina, 174f, 174t
R groups, 40, 41f
Rh factor, 98, 99, 224–225
Ribonucleic acid (RNA), 231, 235–236, 235f
Ribose, 235, 235f

Ribosomal RNA, 231, 236
Ribosomes, 50f, 238, 239, 239f
Ribs, 152f, 156, 156f
Right atrioventricular valve, 93, 93f
Right atrium, 92, 92f, 93f
Right common carotid artery, 82f, 84f
Right coronary artery, 92f
Right pleural cavity, 83
Right pulmonary artery, 92, 92f
Right pulmonary veins, 92f
Right ventricle, 92, 92f, 93f
RNA polymerase, 237
Rod cells, 174t, 175
Root hair plexuses, 178f
Rotating lenses (microscope), 22, 23
Rotation, 160, 160f
Rotifers, 266f
Rough ER, 50f
Round window, 177f
Ruffini endings, 178f

S

Saccule, 177, 177t
Sacrum, 152f, 155, 155f
Sagittal crest, 260, 260f
Sahelanthropus tchadensis, 256f
Sample sizes, 5
SA node, 94, 94f
Sarcolemma, 163f, 166f
Sarcomeres, 163, 163f
Sarcoplasm, 163f
Sarcoplasmic reticulum, 163f
Sartorius, 159f
Saturated fatty acids, 43, 130
Scanning electron microscopes, 19, 20f, 21
Scanning objective, 24
Scapulae, 149f, 152f, 156f, 157, 157f
Scientific method, 1–7, 2f
Sclera, 174f, 174t
Scrotal sac, 78f, 87
Scrotum, 184f
Sea level changes, 272
Sea stars, 188, 188f
Secondary follicles, 186, 187f
Secondary oocytes, 186, 187, 187f, 212, 213f
Seedless vascular plants, 250t
Seeds, acid deposition effects, 268
Selectively permeable membranes, 53, 54–55
Semen, 184
Semicircular canals, 177, 177f, 177t
Semiconservatism, 234
Seminal vesicles, 88, 88f, 184, 184f
Seminiferous tubules, 185, 185f
Semisolids, diffusion in, 52, 52f
Sense organs, 63f
Senses, brain centers for, 171f
Sensory neurons, 172, 173f
Sensory receptors
    basic functions, 167
    in human skin, 178–179, 178f
    for reflexes, 173
    signal conduction, 168, 172, 173f
    sound, 177–178
    taste and smell, 180
Septa, heart, 92, 93f
Serosa, intestinal, 72, 72f

Serratus anterior, 159f
Sertoli cells, 185
Serum, 262–263
Sewage systems, 267, 271
Sex chromosomes, 222–223, 226
Sex differences, 158, 184–187, 222
Sex selection, 229
Sexually transmitted diseases, 32
Sexual reproduction, 199. See also
    Meiosis; Reproductive
    system
Shark fins, 253f, 254
Sheep brain, 168–170, 169f
Sheep heart dissection, 92f, 93, 93f
Shoulder bones, 157
Sickle-cell disease, 243–245, 243f
Sickle-cell trait, 243
Silurian period, 250t
Simple columnar epithelium, 64, 64f
Simple cuboidal epithelium, 64, 64f
Simple epithelium, defined, 62
Simple squamous epithelium, 62, 62f, 63, 100, 139f
Sister chromatids, 199, 200, 201f, 205
Skeletal muscle. See also Muscles
    contractions and movements, 160–164
    naming, 159–160
    superficial, 159f
    tissues, 69, 69f, 162–164, 162f, 163f
Skeleton
    basic anatomy, 152–158, 152f–158f
    human versus chimpanzee, 258–260, 258f, 260f
Skin-fold measurements, 126
Skin receptors, 178–179, 178f
Skin tissues, 62–63, 63f
Skull
    bones, 152f, 153–154
    hominids compared, 256–257
    human versus chimpanzee, 259–260, 260f
Small intestine
    epithelial tissue, 64f
    fetal pig, 84f, 85–86
    functions, 77, 106f, 112, 112f
    location, 76f
    relation to liver, 143f
Smell receptors, 180
Smoking, 148, 214
Smooth ER, 50f
Smooth muscle, 70, 70f
Snails, 251f
Snakes, 252f
Sodium chloride, 56–57
Soft palate, 76f, 79, 79f
Solutions, 52, 55–57
Somatosensory association area, 171f
Somites, 193, 193f, 255
Sound location, 178
Species composition and diversity, in hay infusions, 266–268
Specific dynamic action, 115, 123
Spectrophotometers, 270–271
Speed of diffusion, 52–53
Sperm, 185, 185f, 212, 213f
Spermatic arteries, 88f
Spermatic cord, 88f

Spermatogenesis, 185, 212, 213f
Sphenoid bone, 153, 153f, 154
Sphincters, precapillary, 99, 99f
Spinal column. See Vertebral column
Spinal cord
    anatomy, 172, 172f, 173f
    in central nervous system, 167, 167f
    connection to brain, 169f, 171f
    embryonic, 192, 255f
    in vertebral column, 155
Spinal nerves, 167, 167f, 172, 173f
Spinal reflexes, 172–173
Spindle fibers, 200, 202f–203f
Spindles, 200, 201f, 201t, 202f, 203
Spindle-shaped cells, 70
Spinous processes, 155f
Spiny anteater, 253, 253f
Spirometers, 141–142, 141f
Spleen, 76f, 77, 85
Spongy bone, 150f, 151
S stage, 200, 200f
Stage clips, 25
Stage of microscope, 25
Stapes, 177, 177f
Starch
    digestion, 36, 111–112
    reaction to Benedict's reagent, 39t
    structure, 36f
    testing for, 37–38
Stentor, 266f
Stereocilia, 177t
Stereomicroscopes, 19, 20, 21–23, 22f
Sternum, 152f, 156f
Stethoscopes, 95
St. Martin, Alexis, 105
Stomach
    actions of pepsin, 107–108, 107f
    fetal pig, 85
    functions, 77, 106f
    location, 76f
    relation to liver, 143f
    surgical treatment, 113
Straight hairline, 218f, 219t
Stratified epithelium, defined, 62
Stratified squamous epithelium, 62, 63, 63f
Stress tests, 7
Stretch receptors, 173
Striated muscle, 69, 69f, 162
Structural proteins, 40
Styloid process, 153f
Subclavian artery, 102f
Subclavian vein, 102f
Subcutaneous layer, 63
Subject variability, 5
Submucosa, intestinal, 72, 72f
Substrates, 58, 58f
Subway, 118t
Sucrose, 36
Sugar glider, 253f, 254
Sugar-phosphate backbones, 232f
Sugars
    absorption, 112, 129
    in DNA, 232f, 233
    in foods, 36
    in RNA, 235, 235f
    structure, 36f
    testing for, 39–40, 111
Sulci, 168, 171f

...ava, 92, 92f,
...ges, 154, 260, 260f
...igaments, 174f
...53
...lands, 63f
...psis, 205
...naptic cleft, 166f
Synaptic vesicles, 166f
Syndromes, 212
Synthesis reactions, 36f, 41f, 58, 58f
Systemic capillaries, 101f, 140f
Systemic circuit, 100–101, 101f, 102f
Systole, 95
Systolic pressure, 103

T

Taco Bell, 118t
Tail, embryonic, 255, 255f
Tail, fetal pig, 81f
Talus, 158f
Tarsals, 152f, 158, 158f
Taste receptors, 180
Tay–Sachs disease, 222
Teeth, 79, 257t, 260, 260f
Telophase, 203f, 204, 211f
Telophase I, 205, 207f, 211f
Telophase II, 207f, 209
Temperature and enzyme actions, 58
Temperature receptors, 178f, 179
Temperature scales, 16, 16f
Template strand, RNA, 237, 237f
Temporal bones, 152f, 153, 153f, 154, 154f
Temporal lobe, 168, 169f, 171f
Tertiary period, 249t
Testes
    anatomy and physiology, 184, 184f, 185, 185f
    fetal pig, 87, 88f
Testosterone, 184, 185
Tetrahymena, 266f
Thalamus, 168, 169f
Theories, 3
Thermal effect of food, 115
Thermal pollution, 265, 267
Third ventricle, 169f, 171f
Thoracic cage, 152f
Thoracic cavity, fetal pig, 80, 81f, 82f, 83, 84f
Thoracic curvature, 155f
Thoracic vertebrae, 155, 155f, 156f
Thymine, 233, 235, 236
Thymus gland
    embryonic development, 255
    fetal pig, 82, 82f, 83
    location, 76f
    lymphatic functions, 77
Thyroid gland
    fetal pig, 82, 82f, 83, 84f
    hormones, 77
    location, 76f
Tibia, 152f, 158, 158f
Tibialis anterior, 159f
Tibial tuberosity, 158f
Tidal volume, 141–142

Tissue engineering, 73
Tissue fluid, 131
Tissues
    connective, 66–69, 66f–69f
    defined, 61
    developmental stages, 190–191
    epithelial, 62–65, 62f–65f
    muscle, 69–70, 69f–70f
Tobacco smoke, 148, 214
Tongue, 76f, 78f, 79, 79f
Tonicity, 55–57, 56f
Tonsils, 255
Total magnification, 27
Touch receptors, 178f, 179
Trabeculae, 150f, 151
Trachea
    epithelial tissue in humans, 65f
    fetal pig, 80, 80f, 82, 82f, 83
    functions, 77
    location, 76f
Transcription, 236, 237, 240f
Trans fats, 43, 130
Transfer RNA, 231, 236, 238–239, 238f, 239f
Translation, 236, 238–239, 240f
Translocations, chromosome, 212
Transmission electron microscope, 19, 20f, 21, 21t
Transplants
    bone marrow, 165
    engineered tissues, 73
    fetal pig tissues, 75
    liver, 147
Transport proteins, 40
Transverse processes, 155f
Trapezius, 159f
Triassic period, 249t
Triceps brachii, 149–150, 149f, 159f
Triceps skin fold, 126
Tricuspid valve, 93, 93f
Trilobites, 251t
Triple beam balance, 13–14, 14f
Triplet code of DNA, 238
Trisomy 21, 225
Trochlea, 157f
Trophoblasts, 189, 189f
True ribs, 156f
T tubules, 163f
Tubular reabsorption, 133, 134f, 136
Tubular secretion, 134, 134f, 136
Turner syndrome, 226
T wave, 94, 94f
Twins, 189
Tympanic membrane, 177, 177f, 177t
Type 1 diabetes, 138, 144
Type 2 diabetes, 129, 138, 144

U

Ulna, 149f, 152f, 157, 157f
Umbilical artery, 84f, 87f, 88f, 195
Umbilical cord
    around baby's neck, 198
    fetal pig, 78–79, 78f, 81f, 84f, 87f
    human, 195, 196f
Umbilical vein, 81, 84f, 195

Unattached earlobes, 217f, 218f, 219t
Underweight, 125
Unknowns, testing, 46
Unsaturated fatty acids, 43
Upper limb bones, 157
Uracil, 235, 235f
Urea, 136, 143–144, 146
Ureters, 86, 87f, 88f, 184f
Urethra
    female, 186f
    fetal pig, 86, 87, 87f, 88, 88f
    functions, 77
    location, 76f
    male, 184, 184f
Urinalysis, 137–138
Urinary bladder
    from engineered tissues, 73
    in female, 186f
    fetal pig, 84f, 86, 87f, 88f
    functions, 77
    location, 76f
    in male, 184f
Urinary system
    coordination with other systems, 89f
    fetal pig, 86, 87f
    organs, 76f, 77
Urine, 135–138
Urogenital opening, fetal pig, 78f, 79
Urogenital papilla, fetal pig, 78f, 79, 87f
Urogenital sinus, fetal pig, 87, 87f
Uterine horns, 87, 87f
Uterus, 87, 87f, 186, 186f
Utricle, 177, 177t
Uvula, 79

V

Vacuoles, 67
Vagina, 87, 87f, 186, 186f
Valine, 239t, 244f
Valves (heart), 92, 93, 93f, 95
Variables, experimental, 4–5
Vas deferens
    anatomy and physiology, 184, 184f, 185f
    fetal pig, 87, 88, 88f
Vasectomy, 185
Vegetables, 117t
Veins
    anatomy, 99f
    coronary, 92, 92f
    functions, 99
    liver, 143, 143f, 144, 145
    major, 102f
    pulmonary and systemic, 101f
Venae cavae, 92, 92f, 101f, 102f
Venous system, fetal pig, 80
Ventral, defined, 77
Ventral horn, 173f
Ventral root, 173f
Ventricles (brain), 168, 169f, 171f
Ventricles (heart), 92, 92f, 93f
Venules, 99, 99f, 100, 134f, 135, 135t
Vernix caseosa, 197

Vertebrae, 155, 155f
Vertebral column
    in axial skeleton, 152f
    embryonic development, 192
    human versus chimpanzee, 258, 258f
    structure, 155
Vertebrates
    defined, 251
    evolution, 248f
    fossils, 252–253
    homologous forelimb structures, 254–255, 254f
Vesicles, 50f, 51, 166f
Vesicular follicles, 187, 187f
Vestibular nerve, 177f
Villi, chorionic, 195, 196f
Villi, intestinal, 112, 112f
Visceral muscle, 70
Vision, 174–176
Vision disorders, 222–223
Visual association area, 171f
Vital capacity, 141, 142
Vitelline blood vessels, 193, 193f
Vitreous humor, 174f, 174t
Volume measurement, 14–15
Voluntary muscle, 69
Vomer bone, 153f
Vorticella, 266f

W

Water as polar molecule, 44
Water intoxication, 59
Weight, ideal, 124–129
Weight loss methods, 113, 129–130
Weight measurement, 13–14
Wet mounts, 30, 30f, 31
White blood cells
    basic features, 97, 97f
    as connective tissue, 69
    microscopic observations, 30, 69f, 98
    photomicrographs, 20f, 97f
    in urine, 138
Whitefish blastula, 202
White matter, 172, 172f, 173f
Widow's peak, 218f, 219t
Wrist, fetal pig, 78f, 81f

X

Xiphoid process, 156f
X-linked disorders, 222–223, 223f

Y

Yellow marrow, 150f, 151
Yolk plugs, 190f
Yolk sacs, 191f, 195, 196f
Yolk stalks, 255f

Z

Z line, 163f
Zoom mechanisms, 22, 23
Zooplankton, 272
Zygomatic bone, 152f, 153f, 154, 154f
Zygotes, 183, 188, 188f, 213f